Why Johnny Can't Add:
The Failure of the New

Why Johnny Can't Add: The Failure of the New Math

MORRIS KLINE

PROFESSOR OF MATHEMATICS
COURANT INSTITUTE
OF MATHEMATICAL SCIENCES
NEW YORK UNIVERSITY

VINTAGE BOOKS
A Division of Random House
New York

372.73
K65w

First Vintage Books Edition, February 1974
Copyright © 1973 by Morris Kline

Library of Congress Cataloging in Publication Data
Kline, Morris, 1908–
  Why Johnny can't add: the failure of the new math.
  Bibliography: p.
  1. Mathematics—Study and teaching—United States. I. Title.
[QA13.K62 1974]   372.7′3′044   73–14606
ISBN 0–394–71981–6

Is it life, I ask, is it even prudence,
To bore thyself and bore the students?

Johann Wolfgang von Goethe

Table of Contents

Preface

For many generations the United States maintained a rather fixed mathematics curriculum at the elementary and high school levels. This curriculum, which we shall refer to as the traditional one, is still taught in fifty to sixty per cent of the American schools. During the past fifteen years a new curriculum for the elementary and high schools has been fashioned and has gained rather wide acceptance. It is called the modern mathematics or new mathematics curriculum. Though many groups have contributed to it and their recommendations are not quite identical, for present purposes I believe that it is proper and fair to overlook the differences among them.

The experimental work on the new curriculum, such as it was, has been done. Hundreds of new texts have been written and millions of children and young people have been and are being taught with this new material. In addition, several dozen books have been published which explain the new curriculum to parents, teachers, principals, superintendents, and other interested parties. The money, time, energy, and thought expended on this program have been considerable—even enormous.

Mathematics occupies a central position in the schools.

Students spend eight years on it in the elementary schools and from two to four years in the high schools. Moreover, the subject has proved to be an obstacle to scholastic achievement for many students. Hence the question of whether the new curriculum has actually improved the teaching of mathematics and has indeed made the subject more accessible to the students is important.

Now that the new program has been somewhat stabilized and its nature made clear, it seems possible and necessary to decide whether progress has actually been made. Are our children really better off by reason of this nationwide, highly touted reform? Certainly the education of our children is too important for us to accept a curriculum uncritically just because it has been extensively promoted and has been backed by many professors of mathematics.

Up to the present time it has been assumed by the general public that the profession has spoken and that the only problem is how to extend the teaching of the new curriculum to more and more schools. However, sharp differences of opinion as to the merits of the innovations do exist among professional mathematicians and teachers. It behooves all interested parties to examine the effectiveness of the new material lest the innovations become established as the new orthodoxy despite the absence of any firm evidence that the innovations are genuine improvements. This is what I propose to do.

I hope that the reader will feel, as I do, that any book critical of a particular attempt at reform is not *ipso facto* reactionary. The traditional curriculum has major defects and I shall cite them. It needs to be improved. But it seems to me that true progress is possible—and a truly progressive attitude can exist—only if we have the courage to admit that any particular attempt at reform has not worked.

I am indebted to many people for helpful criticisms

and suggestions but especially to Professor Fred V. Pohle of Adelphi University, Professor Alexander Calandra of Washington University, and to Dr. George Grossman, Director of Mathematics for the New York City Board of Education. I am also indebted to Mr. Thomas McCormack, president of St. Martin's Press, not only for criticisms and suggestions, but for his encouragement to publish this book. He stressed repeatedly that a critique of the mathematics curriculum would be a public service. Of course the particular views expressed herein are chargeable only to myself.

Morris Kline

1973

Why Johnny Can't Add:
The Failure of the New Math

A Taste of Modern Mathematics

". . . Great God! I'd rather be
A Pagan suckled in a creed outworn;
So might I, . . .
Have glimpses that would make me less forlorn."

William Wordsworth

Let us look into a modern mathematics classroom. The teacher asks, "Why is $2 + 3 = 3 + 2$?"

Unhesitatingly the students reply, "Because both equal 5."

"No," reproves the teacher, "the correct answer is because the commutative law of addition holds." Her next question is, "Why is $9 + 2 = 11$?"

Again the students respond at once: "9 and 1 are 10 and 1 more is 11."

"Wrong," the teacher exclaims. "The correct answer is that by the definition of 2,

$$9 + 2 = 9 + (1 + 1).$$

But because the associative law of addition holds,

$$9 + (1 + 1) = (9 + 1) + 1.$$

Now $9 + 1$ is 10 by the definition of 10 and $10 + 1$ is 11 by the definition of 11."

Evidently the class is not doing too well and so the teacher tries a simpler question. "Is 7 a number?" The students, taken aback by the simplicity of the question, hardly deem it necessary to answer; but the sheer habit of obedience causes them to reply affirmatively. The teacher is aghast. "If I asked you who you are, what would you say?"

The students are now wary of replying, but one more courageous youngster does do so: "I am Robert Smith."

The teacher looks incredulous and says chidingly, "You mean that you are the name Robert Smith? Of course not. You are a person and your name is Robert Smith. Now let us get back to my original question: Is 7 a number? Of course not! It is the *name* of a number. $5 + 2$, $6 + 1$, and $8 - 1$ are names for the same number. The symbol 7 is a *numeral* for the number."

The teacher sees that the students do not appreciate the distinction and so she tries another tack. "Is the number 3 half of the number 8?" she asks. Then she answers her own question: "Of course not! But the numeral 3 is half of the numeral 8, the right half."

The students are now bursting to ask, "What then is a number?" However, they are so discouraged by the wrong answers they have given that they no longer have the heart to voice the question. This is extremely fortunate for the teacher, because to explain what a number really is would be beyond her capacity and certainly beyond the capacity of the students to understand it. And so thereafter the students are careful to say that 7 is a numeral, not a number. Just what a number is they never find out.

The teacher is not fazed by the pupils' poor answers. She asks, "How can we express properly the whole numbers between 6 and 9?"

"Why," one pupil answers, "just 7 and 8."

"No," the teacher replies. "It is the set of numbers which is the intersection of the set of whole numbers larger than 6 and the set of whole numbers less than 9."

Thus are students taught the use of sets and, presumably, precision.

A teacher thoroughly convinced of the vaunted value of precise language, and wishing to ask her students whether a number of lollipops equals a number of girls, phrases the question thus: "Find out if the set of lollipops is in one-to-one correspondence with the set of girls." Needless to say, she gets no answer from the students.

Bent but not broken, the teacher asks one more question: "How much is 2 divided by 4?"

A bright student says unhesitatingly, "Minus 2."

"How did you get that result?" asks the teacher.

"Well," says the student, "you have taught us that division is repeated subtraction. I subtracted 4 from 2 and got minus 2."

It would seem that the poor children would deserve some relaxation after school, but parents anxious to know what progress their children are making also query them. One parent asked his eight-year-old child, "How much is 5 + 3?" The answer he received was that 5 + 3 = 3 + 5 by the commutative law. Flabbergasted, he rephrased the question: "But how many apples are 5 apples and 3 apples?"

The child didn't quite understand that "and" means "plus" and so he asked, "Do you mean 5 apples plus 3 apples?"

The parent hastened to say yes and waited expectantly.

"Oh," said the child, "it doesn't matter whether you are talking about apples, pears or books; $5 + 3 = 3 + 5$ in every case."

Another father, concerned about how his young son was getting along in arithmetic, asked him how he was faring.

"Not so well," the boy replied. "The teacher keeps talking about associative, commutative and distributive laws. I just add and get the right answer, but she doesn't like that."

These minor examples may illustrate, and perhaps caricature, some features of the curriculum now called modern mathematics or the new mathematics. We shall examine the major features in greater detail in due course and we shall consider their merits and demerits. But first, we shall review briefly the "old" mathematics to see what defects prompted the development of a new curriculum.

2

The Traditional Curriculum

"I have found you an argument but I am not obliged
to find you an understanding."

Samuel Johnson

Though the traditional curriculum has been affected
somewhat in recent years by the spirit of reform, its basic
features are readily described. The first six grades of
the elementary school are devoted to arithmetic. In the
seventh and eighth grades the students take up a bit of
algebra and simple facts of geometry such as formulas
for area and volume of common figures. The first year
of high school is concerned with elementary algebra, the
second with deductive geometry, and the third with more
algebra (generally called intermediate algebra) and with
trigonometry. The fourth high school year usually covers
solid geometry and advanced algebra; however, there has
not been as much uniformity about fourth-year work as
there has been for the earlier years.

Several serious criticisms of this curriculum have been
voiced repeatedly. The first major criticism, which applies to algebra in particular, is that it presents mechani-

cal processes and therefore forces the student to rely upon memorization rather than understanding.

The nature of such mechanical processes can readily be illustrated. Let us consider an arithmetical example. To add the fractions 5/4 and 2/3, that is, to calculate

$$\frac{5}{4} + \frac{2}{3},$$

students are told to find first the least common denominator, that is, the smallest number into which 4 and 3 divide evenly. This number is 12. One then divides 4 into 12, obtains 3, and multiplies the numerator 5 of the first fraction by 3. Similarly one divides 3 into 12, obtains 4, and multiplies the numerator 2 of the second fraction by 4. The result thus far is to convert the above sum into the equal sum

$$\frac{15}{12} + \frac{8}{12}.$$

One now sees easily that the sum is 23/12.

A good teacher would no doubt do his best to help students grasp the rationale of this process, but on the whole the traditional curriculum does not pay much attention to understanding. It relies upon drill to get students to do the process readily.

After students learn to add numerical fractions they face a new hurdle when asked in algebra to add fractions where letters are involved. Though the same process is used to calculate

$$\frac{3}{x + a} + \frac{2}{x - a}$$

the individual steps are more complicated. Again the curriculum relies upon drill to put the lesson across. The students are asked to carry out the additions in numerous exercises until they can perform them readily.

They are taught many dozens of such processes: factoring, solving equations in one and two unknowns, the uses of exponents, addition, subtraction, multiplication and division of polynomials, and operations with negative numbers and radicals such as $\sqrt{3}$. In each case they are asked to imitate what the teacher and the text show them how to do. Hence the students are faced with a bewildering variety of processes which they repeat by rote in order to master them. The learning is almost always sheer memorization.

It is also true that the various processes are disconnected, at least as usually presented. They rarely have much to do with each other. While all these processes do contribute to the goal of enabling the student to perform algebraic operations in advanced mathematics, as far as the students can see the topics are unrelated. They are like pages torn from a hundred different books, no one of which conveys the life, meaning and spirit of mathematics. This presentation of algebra begins nowhere and ends nowhere.

After a year of such work in algebra the traditional curriculum shifts to Euclidean geometry. Here mathematics suddenly becomes deductive. That is, the text starts with definitions of the geometrical figures and with axioms or basic assertions which are presumably "obviously true" about the figures. They then prove theorems by applying deductive reasoning to the axioms. The theorems follow each other in a logical sequence; that is, the proofs of later theorems depend upon the conclusions already established in the earlier theorems. The sudden shift from mechanical algebra to deductive geometry certainly bothers most students. They have not thus far in their mathematics education learned what "proof" is and must master this concept in addition to learning subject matter proper.

The concept of proof is fundamental in mathematics, and so in geometry the students have the opportunity to learn one of the great features of the subject. But since the final deductive proof of a theorem is usually the end result of a lot of guessing and experimenting and often depends on an ingenious scheme which permits proving the theorem in the proper logical sequence, the proof is not necessarily a natural one, that is, one which would suggest itself readily to the adolescent mind. Moreover, the deductive argument gives no insight into the difficulties that were overcome in the original creation of the proof. Hence the student cannot see the rationale and he does the same thing in geometry that he does in algebra. He memorizes the proof.

Another problem troubles many students. Since algebra is also part of mathematics, why is deductive proof required in geometry but not in algebra? This problem becomes more pointed when students take intermediate algebra, usually after the geometry course, because there proof is again abandoned in favor of techniques.

With or without proof, the traditional method of teaching results in far too much of only one kind of learning—memorization. The claim that such a presentation teaches thinking is grossly exaggerated. By way of evidence, if evidence is needed, I have challenged hundreds of high school and college teachers to give open book examinations. This suggestion shocks them. But if we are really teaching thinking and not memorization, what could the students take from the books?

The traditional curriculum has also become too traditional. Some topics that received considerable emphasis for generations have lost significance but are still retained. One example is the solution of triangles in trigonometry. Here, given some parts—sides and angles—of a triangle, the theory shows how to compute other parts and even how to use logarithms in the calculations. This topic, which had far more relevance when trigonometry was taught primarily to prospective surveyors, should have been deemphasized long ago. Another example is the computation of irrational roots of polynomial equations. The method usually taught, called Horner's method, requires several weeks of class time and does not warrant it.

There are also minor logical defects in the traditional curriculum. For example, students are taught that $x^2 - 4$ can be factored into $(x + 2)(x - 2)$, but that $x^2 - 2$ cannot be factored. However, the latter can be factored if we are willing to introduce irrational numbers. In this event the factors are $x - \sqrt{2}$ and $x + \sqrt{2}$. Likewise $x^2 + 4$ can be factored if we are willing to use complex numbers. In this case the factors are $x + 2i$ and $x - 2i$ where $i = \sqrt{-1}$. Thus the error made in the traditional method of teaching is the failure to specify the class of numbers we are willing to consider in order to perform the factoring.

Beyond the few defects we have already described, the traditional curriculum suffers from the gravest defect that one can charge to any curriculum—lack of motivation. Mathematics proper, to use the words of the famous twentieth-century mathematician Hermann Weyl, has the inhuman quality of starlight, brilliant and sharp, but cold. It is also abstract. It deals with mental concepts, though some, such as geometrical ones, can be visualized. On both accounts, the coldness and the abstractness, very few students are attracted to the subject.

Young people can no doubt see that there is some point to learning arithmetic but they can see little reason to study algebra, geometry and trigonometry. Why should they learn the addition of algebraic fractions, the solution of equations, factoring and other topics? The appeal of geometry is not greater. It is true that students can see what geometry is about and what the theorems assert; the figures make clear what this branch of mathematics deals with. But the question of why one should study this material is still not answered. One can readily understand what the history of China is about, but may still question why he should be obliged to learn it. Why is it important to know that the opposite angles of a parallelogram are equal or that the altitudes of a triangle meet in a point?

Clearly one cannot defend algebra, geometry and trigonometry on the ground that they will be of use later in life. The educated layman does not have occasion to use this knowledge at any time unless he becomes a professional scientist, mathematician, or engineer. But this group cannot be more than a few per cent of the high school population. Moreover, even if all of the students were to use some mathematics later in life this usage cannot be motivation. Young people cannot be asked to take seriously material that they might need years later. This motivation is often described as offering "pie in the sky."

As a matter of fact, in an effort to motivate the students, the schools did try to teach some uses of arithmetic in the seventh and eighth grades. They taught simple and compound interest and discount on loans. But twelve- and thirteen-year-old students did not take to such material and the experiment is conceded to be a failure. The motivation must appeal to the student at the time he takes the course.

Another motivation often dangled before students is that they must study mathematics to get into college. If the mathematics they have been taught in elementary and high school is typical of what lies ahead in college, they may not want to go to college.

The prospective mathematicians, scientists, and engineers will find mathematics useful in their careers. But if the mathematics presented gives no inkling of how it will be useful and if it is in itself totally unattractive, telling the students that it is needed in science and engineering will only encourage them to seek another career.

Much of the mathematics taught is often defended as "training the mind." There may very well be some training, but the same effect can be achieved with subject matter that is far more understandable and agreeable. One could teach the commonly used forms of reasoning by resorting to social or simple legal problems whose relevance to life is far more apparent to the students. One does not need mathematics to teach people that the statement "All good cars are expensive" is not the same statement as "All expensive cars are good." Moreover, the use of social or legal problems does not require the mastery of technical language, symbolism, and abstract concepts, which tend to obscure the reasoning. Thus it is far more difficult for the student to see that the statement "All parallelograms are quadrilaterals" is not the same as "All quadrilaterals are parallelograms." In fact, experience in teaching shows that to make the logical arguments used in mathematical reasoning clear to the student, one must use nonmathematical examples involving the same arguments. Moreover, there is some question about whether the training to think in one sphere carries over to thinking in another. One may be inclined to believe that it does, but one could not prove that this is so.

Another commonly advanced justification for teaching mathematics at the high school level is the beauty of the subject. But we know that the subjects taught have not been selected because they are beautiful. They have been selected because they are necessary for further work in mathematics. There is no beauty in adding fractions, in the quadratic formula, or in the law of sines. No amount of preaching or rhapsodizing about the beauty of mathematics will make such ugly ducklings appealing. Moreover, novitiates are not likely to find beauty in a subject they are still striving to master, any more than one who is learning French grammar can appreciate the beauty in French literature.

A few students are attracted to mathematics by the intellectual challenge or because they like what they happen to do well. The rare student who experiences this challenge may indeed be intrigued—as some mathematicians are—by the fact that there are only five regular polyhedra. However, as far as most students are concerned, the world would be just as well off if there were an infinite number of them. As a matter of fact, there is an infinite number of regular polygons and no one seems depressed by this fact.

There is indeed an intellectual value in mathematics. But there is a question of whether young people can appreciate it just as there is about whether a six-year-old can appreciate Beethoven's music. If the teacher proves a theorem of mathematics, the student will still be struggling to understand the theorem, its proof and its meaning. While undergoing such struggles the student is not likely to be impressed with the intellectual content and what the human mind has accomplished. In him the theorem and proof produce bewilderment and confusion.

Beyond the purported values of training the mind, beauty and intellectual challenge, the defenders of the traditional curriculum point to the exercises. These, they say, show uses of mathematics and should convince the student that the material is important. There are work problems such as the ditch-digger's dilemma. "One man can dig a ditch in two days and another in three days. How much time will be required if both men dig it together?" Such problems create pointless work.

Then there are tank-filling problems for students who

have no swimming pools to fill. Or the mixture problems: "How many quarts of milk with ten per cent cream and how many quarts of milk with five per cent cream must be mixed to make a hundred quarts of milk with fifty per cent cream?" Such problems are useful to farmers who wish to fake the cream content of their milk. Other mixture problems concern mixing brands of coffee or brands of tea to make undrinkable brews.

There are age problems too: "Jane is twenty years older than Mary. In ten years Jane will be twice as old. How old is Mary?" This type of problem calls for finding out other people's ages, and many people are sensitive about their ages.

There are also number problems, such as "One number is three times another number minus two. What are the numbers?" (The numbers racket is actually illegal.) More "realistic" are board problems. "A board seven feet long is to be cut into two parts, one of which is to be two feet longer than the other. How long are the parts?" Of course students are bored with board problems.

And we shouldn't neglect to mention the time, rate and distance problems, such as up- and down-river travel for students who are going nowhere and whose desire to go anywhere has not been aroused. Some problems involve taking walks around a circular garden and ask for the dimensions of the garden. If we allowed the students to take walks around the garden and provided each with a pleasant companion we would do the students more good.

All these problems are hopelessly artificial and will not convince anyone that algebra is useful.

Some authors of algebra texts do point to "truly physical" problems. For example, Ohm's law states that the voltage E equals the current I times the resistance R. In

symbols $E = IR$. Calculate E if $I = 20$ and $R = 30$.
But the current involved in such problems doesn't drive
any mental motors. So far as the student is informed
Ohm's law could be describing the number of marriages
in Burma each year.

For generations the calculus textbooks have asked
students to calculate centers of gravity and moments of
inertia of bodies without ever pointing out why these
quantities are significant. Consequently, the gravity of
these problems produces nothing but inertia in the students. Such physical problems, presented with no pre-
liminary explanation of physical background or physical
significance, mean nothing to the student. Clearly, a
physical application is worthless if the student cannot see
what is accomplished.

Even the use of the word "application" is often bother-
some. Students are taught, say, a formula for area and
are then asked to calculate areas with it. These calculations are supposed to be an "application." This kind
of application adds insult to injury. Since the so-called
applications are still pointless and still part of mathematics proper, in what sense are they applications?

The fact is, then, that no motivation for the study of
mathematics is offered in the traditional curriculum.
Students take it because they are required to. Motivation
means more than a psychological stimulus. Genuine motivation also supplies insight into the very meaning of
the mathematics. A great deal of mathematics, particularly on the elementary level, was suggested directly by
real situations and problems. The bare formula $s = 16t^2$
acquires meaning when one learns that it relates the
distance fallen and time of travel of an object which is
dropped. An ellipse becomes more than just another
curve when one learns that it is the path of a planet
around the sun. Moreover, the questions that are raised
about the formula and about the curve become mean-

ingful because they concern the physical situations. The physical meanings also supply, in many cases at least, the power to think about the mathematical problems that are raised, because the mathematics is no more than a representation of the physics and a means of solving physical and other problems.

The failure to present the meaning of mathematics is analogous to teaching students how to read musical notation without allowing them to play the music. Students might be taught how to recognize full notes, half notes, sharps, flats, the key, and how to transpose music from one key to another without ever hearing any music. But if they do not hear what these various notations and techniques mean, they will be left with meaningless and boring skills.

The traditional curriculum has been faithfully reproduced in thousands of textbooks. The strongest reaction induced by the traditional texts is that they are insufferably dull. Most textbook writers seem to believe that scientific writing must be cold, spiritless, mechanical and dry. These books have no authors. They are not only printed by machines; they are written by machines.

Textbook writers also seem to take inordinate pride in brevity, which can often be interpreted as incomprehensibility. Reasons for steps are either not given or given so briefly as to make the presentation almost useless to the student. Many authors seem to be saying, "I have learned this material and now I defy you to learn it." Brevity in mathematical exposition is the soul of witlessness and obscurity.

The most disturbing fact about many traditional mathematics texts is that they lack originality and repeat each other endlessly. A few thousand arithmetic, algebra, geometry, and trigonometry texts have been published since 1900. Practically all of the texts on any one of these subjects contain the same material and presentation; only the order of the topics is different. But there is hope for "progress" because each contains at least ten topics, and the number of permutations of ten objects ten at a time is 3,628,800. It would be difficult to estimate how many trigonometry texts have been written with the justification that they treat the general angle before the acute angle. One can be sure, however, that just as many boast of treating the acute angle before the general angle. The only thing that is acute about these books is·the pain they give the reader.

Are there no variations among these books? There are variations such as the elementary algebra and the advanced algebra, the elementary advanced algebra and the advanced elementary algebra, the half-course and the full course, the seven-eighths course, and so forth. Here, too, there is hope for "progress" because there are irrational numbers; hence, we can look forward to irrational algebra courses.

What is especially disturbing about these books is that many of the authors are consciously dishonest to their profession. I asked one professor who had written "umpteen" trigonometries of the full and partially full type why he included such useless topics as the solution of oblique triangles by the law of tangents and the law of half-angles. He admitted that these topics are worthless, but said he included them because the books sell better. Apparently, no matter how many trigonometries a man may write, not even one can reflect his honest judgment.

I asked another professor, who published a stereotyped college algebra, why he bothered to write such repetitious nonsense. "Oh," he said, "I can write the stuff between classes without having to think about it. Why shouldn't I do it?" Needless to say, no thinking was evident in the presentation of the material.

Another professor published a book which included

some material that he believed to be unimportant. He admits this in his preface and then says quite candidly that he included this material with an eye to the market. Such honest dishonesty!

Those authors who repeat each other's topics are in a sense plagiarizing. But the plagiarism extends beyond that. Paraphrases of whole sections of material covering many paragraphs are readily found. One author took whole chapters from another book with only minor changes, acknowledging, of course, the inspiration of God, Euclid, Newton, and Einstein.

Most traditional mathematics textbooks appear to be commercial jobs that make a contribution only to the authors' pocketbooks. The ethics of some teachers, to say nothing of their mentalities, is evidently in a sad state. The only persons who can claim any credit for original work in connection with these books are the publishers' publicity men, who must think up good blurbs for the advertisements.

Fairness requires that one mention recent improvements in the format of mathematics texts. Important formulas are now enclosed in red boxes. Other texts use overlays of plastic to show the increasing complexity of a figure as one overlay after another is superimposed on the original text figure.

Clearly the defects of the traditional curriculum are numerous. The reliance upon memorization of processes and proofs, the disparate treatments of algebra and geometry, minor logical defects, the retention of a few outmoded topics, and the absence of any motivation or appeal explain why young people do not like the subject and therefore do not do well in it. Their dislike is intensified and their difficulties in understanding are compounded by being asked to read dull, poorly written, and commercially contrived textbooks.

Certainly reform was called for. The leaders of the new mathematics movement did not cite all of the above defects. However, they did point the finger at some of them. So let us look now at what these people proposed to do and try to evaluate their effectiveness in improving the teaching of mathematics.

3

The Origin of the Modern Mathematics Movement

"Experience, however, shows that for the majority of the cultured, even of scientists, mathematics remains the science of the incomprehensible."

Alfred Pringsheim

There was general agreement in the early 1950s and even before that date that the teaching of mathematics had been unsuccessful. Student grades in mathematics were far lower than in other subjects. Student dislike and even dread of mathematics were widespread. Educated adults retained almost nothing of the mathematics they were taught and could not perform simple operations with fractions. In fact, these people did not hesitate to say that they got nothing out of their mathematics courses. When this country entered World War II, the military discovered quickly that the men were deficient in mathematics and that they had to institute special courses to bring up the level of proficiency.

Though there are many factors that determine the outcome of any teaching activity, the groups that undertook reform focused on curriculum and argued that if this component were improved the teaching of mathematics would be successful.

In 1952 the University of Illinois Committee on School Mathematics headed by Professor Max Beberman began fashioning a new, or modern, mathematics curriculum. By 1960 the curriculum (at that time directed solely toward the high schools) was used on an experimental basis. Subsequently, the Committee undertook to provide an elementary school curriculum and gradually extended the teaching of both the elementary and high school subject matter to additional geographical areas. The experimental texts, in photo-offset form, were eventually published as commercial texts.

In 1955 The College Entrance Examination Board, whose function is to prepare college entrance examinations which meet the requirements of many colleges, decided to take up the problem of the high school mathematics curriculum and compose what it considered to be the proper one. It set up its own Commission on Mathematics. In 1959 the Commission issued its report, *Program for College Preparatory Mathematics,* and adjoined several appendices which contained samples of recommended subject matter. It did not produce texts. During the years 1955 to 1959 and for several years thereafter, the members of the Commission toured the country and campaigned for the kind of curriculum it proposed in its *Program.*

In the fall of 1957 the Russians launched their first Sputnik. This event convinced our government and country that we must be behind the Russians in mathematics and science and had the effect of loosening the purse strings of governmental agencies and foundations. It may be coincidence but at this time many other groups decided to go into the business of producing a new curriculum.

The American Mathematical Society, the organization concerned with research, decided in 1958 that its talents should be applied to the fashioning of a high school curriculum, and it set up a new group, The School Mathematics Study Group, headed by Professor Edward G. Begle, then at Yale University, to undertake the task. This group began its work by writing curricula for the junior and senior high schools and then extended its program to include the elementary school arithmetic curriculum.

The National Council of Teachers of Mathematics set up its own curriculum committee, The Secondary School Curriculum Committee, which came out with its recommendations in an article in the May 1959 issue of *The Mathematics Teacher*. Many other groups, such as the Ball State Project, the University of Maryland Mathematics Project, the Minnesota School Science and Mathematics Center, and the Greater Cleveland Mathematics Program, were soon formed and began their work.

Individual high school and college teachers commenced in the late 1950s to write their own texts along the lines already foreshadowed or explicitly recommended by the curriculum groups. By the early 1960s a spate of such books had appeared, and many more have continued to appear since that time.

Rather surprisingly, the many groups and independent textbook writers all headed in about the same direction. Hence they have all, fairly enough, been described by the term "modern mathematics" (or "new mathematics").

The origin of the term modern mathematics is relevant. Even before the members of the Commission on Mathematics had determined just what they were going to recommend, they gave addresses to large groups of teachers. Their main message was that mathematics education had failed because the traditional curriculum offered antiquated mathematics, by which they meant mathematics created before 1700. Implicit in this contention was the assumption that young people were aware of this fact and therefore refused to learn the material. Would you, argued these educators, go to a lawyer or a physician whose knowledge of his profession was limited to what was known before 1700? Though these speakers were presumably informed in mathematics they ignored completely the fact that mathematics is a cumulative development and that it is practically impossible to learn the newer creations if one does not know the older ones. Nevertheless, the Commission contended that we must drop the traditional subject matter in favor of such newer fields as abstract algebra, topology, symbolic logic, set theory, and Boolean algebra. The slogan of reform became "modern mathematics."

As it turned out, the reform offered as much a new approach to the traditional curriculum as it did new contents, and some groups emphasized this fact. Hence the term modern mathematics is not really an appropriate description of the new curricula. However, perhaps because the propaganda value of the word modern was too useful to drop—a 1970 automobile is clearly more desirable than a 1969 model—the terms modern mathematics or new mathematics have been retained.

While the modern or new mathematics curriculum as it stands today was being fashioned by the groups already mentioned, new groups appeared on the scene and began to recommend more radical reform. For example, an international group meeting at Royaumont, France, in 1959 urged the abandonment of virtually all the familiar courses in high school mathematics, including Euclidean geometry. The conference declared these subjects to be outdated by electronics, relativity, computers and the soaring importance of abstract mathematics as the basis of modern science. The new subjects were to be logic, structure, and the unity of mathematics as a whole and were to be taught in a new language. This conference did not result in the formation of another curriculum group, but it encouraged still further departures from the traditional curriculum.

Of the newer groups which have proposed more radical reforms we shall mention two. In the summer of 1963 a group of mathematicians assembled for The Cambridge Conference on School Mathematics. (Its report, *Goals for School Mathematics,* appeared as a publication by the Houghton Mifflin Company.) This group recommended the inclusion—by the end of grade twelve, the fourth year of high school—of many additional advanced topics drawn from the theory of numbers, abstract algebra, linear algebra, n-dimensional geometry, projective geometry, tensors, topology, differential equations, and of course, the calculus. On page 7 the report asserts, "The subject matter which we are proposing can be roughly described by saying that a student who has worked through the full thirteen years of mathematics in grades K to 12 (kindergarten through the fourth year of high school) should have a level of training comparable to 3 years of a top-level college training today."

The justification for advocating such a program when the already existing curriculum groups had barely begun to try out their programs or were still fashioning them was given in the Foreword by Francis Keppel, who was then United States Commissioner of Education. He observed that recent curriculum changes are essentially different from those attempted in the past and that the reforms have been eminently successful for the most part (How Dr. Keppel knew this in 1963 when most new mathematics curricula had hardly been tried is not clear.), so much so that "it has sometimes been diffi-

cult to distinguish their shortcomings. Yet the shortcomings are there, and they are by no means insignificant. It can be argued, in fact, that the deficiencies of
the present reform movement are grave enough to
threaten the expressed goals of the movements themselves." Keppel then noted that the changes recommended by the Cambridge group were intended to represent the subject as the scholar saw the discipline, and
that the students were assumed to be able to learn far
more than they had been expected to in the past. The
limitations of the teacher were noted too! "Most curriculum reforms, practically enough, have chosen to limit
their ambitions in the light of these realities. They have
tended to create such new courses as existing teachers,
after enjoying the benefits of brief retraining, can com-
petently handle. They have done so fully aware that
they are thus setting an upper limit, and an upper
limit that is uncomfortably close."

Keppel then continued: "If the matter were to end
there, the result might well be disastrous. New curricula
would be frozen into the educational system that would
come to possess, in time, all the deficiencies of curricula
that are now being swept away. And in all likelihood, the
present enthusiasm for curriculum reform will have long
since been spent; the 'new' curricula might remain in
the system until, like the old, they become not only inadequate but in fact intolerable. Given the relative con-
servatism of the educational system, and the tendency of
the scholar to retreat to his own direct concerns, the lag
may well be at least as long as it has been during the
first half of this century.

"The present report is a bold step toward meeting this
problem. It is characterized by a complete impatience
with the present capacities of the educational system.
It is not only that most teachers will be completely incapable of teaching much of the mathematics set forth

in the curricula proposed here; most teachers would be hard put to comprehend it. No brief period of retraining will suffice. Even the first grade curriculum embodies notions with which the average teacher is totally unfamiliar.

"None the less [*sic*], these are the curricula toward which the schools should be aiming. . . ."

The second of the newer groups joining the movement to revise curricula, the Secondary School Mathematics Curriculum Improvement Study, was organized in 1965 by Professor Howard Fehr of Columbia University. Its goal is to reconstruct secondary school mathematics "from a global point of view." It seeks to eliminate the barriers separating the several branches of mathematics and to unify the subject through its general concepts, sets, operations, mappings, relations and structure. (We shall discuss these concepts later.) Professor Fehr's contention is that his organization of the subject matter will permit the introduction into the high school curriculum of much that has been considered collegiate mathematics. The work of the Cambridge group and of the Curriculum Improvement Study has proceeded slowly and their effect on the schools is not widespread as yet. Hence our account and evaluation of the modern mathematics movement will concentrate on the curriculum efforts of the preceding groups, some of which are still at work on one aspect or another of the school programs.

The curricula which have been formulated by these several organizations are the product of group efforts in which research mathematicians, college and high school teachers and even representatives of industry have collaborated. On the face of it, such collaboration would seem to be a wise procedure. However, attempts to achieve a meeting of minds often result in compromises that are not satisfactory to anyone or which vitiate the thrust of the effort. The point may be illustrated by the story that the famous dancer Isadora Duncan offered herself in marriage to Bernard Shaw and perhaps somewhat facetiously said, ". . . and think of the child who would have your brains and my looks." "Yes," said Shaw, "but what if the child should have your brains and my looks?"

When one seeks to determine what changes these curricula offer, why these changes are desirable, and what reasoning or evidence can be proffered to support the desirability of these changes, one is faced with a problem of considerable magnitude. It is true that in its 1959 report the Commission on Mathematics of the College Entrance Examination Board did describe the contents it recommended. However, except for stressing that modern society requires a totally new mathematics the Commission did not defend the contents it proposed. Moreover, the various curriculum groups that did write texts not only extended the reform to the elementary school grades but did not necessarily follow the Commission's recommendations. One would have expected that each group would have declared its own position and have presented its case for including or excluding particular topics and for adopting its own approach. No such documents have been issued. This is all the more true of the many texts published by individual authors which proclaim themselves to be modern in character. Hence we are left to infer for ourselves what the modern mathematics curriculum is and why it is presumably superior to the traditional curriculum. Could the absence of explanatory and justifying material be interpreted to mean that the advocates of modern mathematics are not too clear themselves on where they have headed, or are they fearful that explicit statements of the features and purported merits of their materials will not bear scrutiny? In any event, to determine the nature and qualities of the modern mathematics curriculum, one must examine the texts and listen to the speeches made by various proponents. At the moment, pending a fuller discussion, let us note that there are two main features of the new curriculum: a new approach to the traditional mathematics, and new contents.

Since we intend to evaluate the new mathematics, it is necessary to consider on what basis one should judge it. One could use as a criterion, Is the mathematics correct? The answer is yes, but the criterion is useless. Correctness does not guarantee that the students will take to the material, that they can absorb it, or that this particular mathematics is what should be taught.

Will it develop mathematicians? Even if it were the ideal curriculum for the training of mathematicians one could not be content. The new mathematics is taught to elementary and high school students who will ultimately enter into the full variety of professions, businesses, technical jobs, and trades, or become primarily wives and mothers. Of the elementary school children, not one in a thousand will be a mathematician; and of the academic high school students, not one in a hundred will be a mathematician. Clearly then, a curriculum that might be ideal for the training of mathematicians would still not be right for these levels of education.

The contents should contribute to the goals of elementary and high school education and should be accessible to young people. The approach to the material should make the content inviting and aid comprehension as far as possible. In particular the new mathematics should remedy at least some of the defects of the traditional curriculum. Unfortunately, in the field of education, unlike mathematics proper, one cannot give an ironclad proof that a particular principle or topic is right or wrong. But there are arguments which do enable us to decide.

Though a dozen or more groups have fashioned new curricula and by now many series of new mathematics texts are on the market, we have already noted that they all adopted about the same approach and contain about the same material. This uniformity has resulted in part from imitation. It is also a consequence of the emphasis and direction which mathematicians are favoring in current research and which we shall discuss at greater length later. Hence, though not every statement we shall make about the new mathematics applies necessarily to any one curriculum, it is fair to treat them as a single movement characterized by common features and content.

We intend to consider carefully the nature of the new mathematics program and to discuss its merits and demerits. Before doing so we should like to inject a somewhat different but nevertheless relevant criticism. Reform of mathematics education was called for, but there is a serious question as to whether curriculum was the weakest component and should have been tackled first. It would, I believe, be generally conceded that the policy of universal education pursued in the United States is highly commendable, but our country was not and still is not prepared to carry on such a program. Certainly we do not have enough qualified teachers; therefore the education in many parts of this country is woefully weak. Were more good teachers available they would have been able long ago, by acting in concert, to remedy the defects of the traditional curriculum. Since the teacher is at least as important as the curriculum, the money, time and energy devoted to curriculum reform might well have been devoted to the improvement of teachers. It is true that in 1958 the National Science Foundation inaugurated and has maintained various institutes for the education of teachers. These institutes should have been used to improve the mathematical backgrounds of elementary and high school teachers so that they could form more independent judgments of what is important in mathematics. Unfortunately, they have been used largely to teach teachers how to teach mathematics of unproven worth.

Whether or not curriculum reform should have received priority, the historical fact must be faced that the new curriculum is at hand and is being widely used. Let us therefore attempt to evaluate it.

4

The Deductive Approach to Mathematics

"The great science [mathematics] occupies itself at least as much with the power of imagination as with the power of logical conclusion."

Johann Friedrich Herbart

One of the major criticisms of the traditional curriculum is that students learn to do mathematics by rote, by memorizing procedures and proofs. It is the contention of the advocates of the modern mathematics curriculum that when the subject is taught logically, when the reasoning behind steps is revealed, students will no longer have to rely upon rote learning. They will *understand* the mathematics. The logical approach is, in other words, also the pedagogical approach and the panacea for the difficulties students have had in learning mathematics.

Just what does the logical approach mean? Basically it is the one commonly used in the traditional curriculum to teach high school geometry. That is, one starts with definitions and axioms and proves conclusions, called theorems, deductively. Though this approach has been used in geometry, it has not been used in the teaching of arithmetic, algebra, and trigonometry. Hence, so far as this feature of the new curriculum is concerned, the major change is in these latter subjects. Let us see what the deductive approach to arithmetic and algebra entails.* The examples we shall examine are taken from typical modern mathematics texts.

The approach to arithmetic usually presupposes that we know what the numbers 0,1,2, . . ., called the counting numbers, are. It then introduces axioms. We know that 4 added to 3 yields the same number as 3 added to 4, or $3 + 4 = 4 + 3$. This axiom is stated in symbolic language as $a + b = b + a$ and is called the *commutative* axiom. That is, we may commute or interchange the order in which we add two numbers.

If we had to calculate $3 + 4 + 5$ we could calculate $3 + 4$ and then add 5 to the result or we could calculate $4 + 5$ and add this result to 3. That is, we could first associate the 3 and 4 and then add 5 or we could first associate the 4 and 5 and add the result, 9, to the 3. Thus we have the *associative* axiom of addition. In symbolic language it states that $(a + b) + c = a + (b + c)$. The commutative and associative axioms apply to multiplication also. That is, $a \times b = b \times a$ and $(a \times b) \times c = a \times (b \times c)$.

Let us consider next $3 \times (4 + 5)$. This is 3×9 or 27. It is also $3 \times 4 + 3 \times 5$. That is, $a \times (b + c) = a \times b + a \times c$. This fact is another axiom and it is called the *distributive* axiom. That is, we may distribute the multiplication over the b and the c instead of applying it to $b + c$.

There are other axioms. For example, the sum and product of two counting numbers is a unique counting

---

* The reader who is familiar with the deductive approach may want to skip the examples in the next few pages.

number. There is a unique number 0, such that $a + 0 = a$ for each counting number a, and there is a unique number 1 such that $a \times 1 = a$ for each counting number a.

These axioms are used to justify steps in arithmetic. Consider the addition of 38 and 3. This is performed as follows. By the definition of 38

$$38 + 3 = (30 + 8) + 3.$$

The associative axiom tells us that

$$(30 + 8) + 3 = 30 + (8 + 3).$$

Now, by the definition of 3,

$$30 + (8 + 3) = 30 + (8 + [2 + 1])$$

and by the associative axiom for addition

$$30 + (8 + [2 + 1]) = 30 + ([8 + 2] + 1).$$

Addition of 8 and 2 yields

$$30 + ([8 + 2] + 1) = 30 + (10 + 1).$$

By the associative axiom

$$30 + (10 + 1) = (30 + 10) + 1,$$

and since $30 + 10 = 40$ and $40 + 1 = 41$, we have finally deduced that

$$38 + 3 = 41.$$

To see how the distributive axiom is used in arithmetic let us consider 7×13. By the definition of 13

$$7 \times 13 = 7 \times (10 + 3).$$

Now, the distributive axiom tells us that

$$7 \times (10 + 3) = 7 \times 10 + 7 \times 3.$$

Since $7 \times 10 = 70$ and $7 \times 3 = 21$, we have deduced that

$$7 \times 13 = 91.$$

Presumably these steps, which employ the distributive axiom, make clearer how the 91 is arrived at than by multiplying in the usual fashion wherein one says that $7 \times 3 = 21$, writes down the 1, carries the 2 over to the tens column and adds it to the result obtained from multiplying 7 and 1.

It is true that our method of writing numbers such as 13 employs what is called positional notation. That is, the 1 in 13 stands for 10 and this must be taken into account in any method of teaching multiplication. Whether citing the distributive axiom clarifies the operation is the issue. Let us suppose that it does and see what it leads to. Consider the problem of 17×13. To follow the above pattern we must do as follows:

$$17 \times 13 = (10 + 7) \times (10 + 3).$$

Now $10 + 7$ must be regarded as a single number and by the distributive axiom

$$(10 + 7) \times (10 + 3)$$
$$= [10 + 7] \times 10 + [10 + 7] \times 3.$$

By the commutative axiom of multiplication

$$[10 + 7] \times 10 + [10 + 7] \times 3$$
$$= 10 \times [10 + 7] + 3 \times [10 + 7],$$

and again by the distributive axiom

$$10 \times [10 + 7] + 3 \times [10 + 7]$$
$$= (10 \times 10 + 10 \times 7) + (3 \times 10 + 3 \times 7).$$

Calculation of the quantities in the parentheses and addition of the two results yield 221. We can readily en-

vision the steps that would have to be undertaken to
multiply, say, 172 by 135.

At some stage in the development of arithmetic and
algebra, usually between the seventh and the ninth
grades, students are asked to learn negative numbers.
To "motivate" the introduction of negative numbers
students are asked first, what number x satisfies the equation $17 + x = 21$? Here the answer is clearly 4. Now
comes the crucial question. What number x satisfies the
equation $21 + x = 17$? To answer this question, one
writes 21 as $17 + 4$ so that

$$(17 + 4) + x = 17.$$

By the associative axiom

$$17 + (4 + x) = 17.$$

But by the definition of 0

$$17 + 0 = 17,$$

so that, since the zero is unique,

$$4 + x = 0.$$

We see that if there were a number x such that $x + 4 =
0$ then we would be able to solve the original equation.
We are therefore motivated to introduce the number -4
with the understanding that $4 + (-4) = 0$.

The counting numbers (except 0) are now called the
positive integers and the new numbers are called the
negative integers. To operate with negative integers the
axioms or basic properties that apply to the counting
numbers are now assumed to hold for the combination
of the old counting numbers and the new negative
numbers. Thus, since the commutative and associative
axioms of addition hold for the counting numbers, they
hold for the positive and negative integers.

For example, to add -2 and -5 we note first that by the definition of -2 and -5,

$$(-2 + 2) + (-5 + 5) = 0 + 0 = 0.$$

But by the commutative and associative axioms

$$(-2 + 2) + (-5 + 5) = 2 + 5 + [(-2) + (-5)],$$

so that

$$0 = 7 + [(-2) + (-5)].$$

Hence $-2 + (-5)$ must be -7 because when it is added to 7 it gives 0.

Having learned how to add negative numbers and positive and negative numbers students are told that subtracting a number means adding its inverse. 4 is the inverse of -4 and conversely. Hence

$$17 - 13 = 17 + (-13) = 4$$
$$6 - 8 = 6 + (-8) = -2$$
$$-5 - (-11) = -5 + 11 = 6.$$

Before establishing additional properties of negative numbers, let us prove that $a \times 0 = 0$ for every integer a. Since by the definition of 0, $a + 0 = a$, then

$$a \times (a + 0) = a \times a.$$

However, by the distributive axiom

$$a \times (a + 0) = a \times a + a \times 0.$$

Hence

$$a \times a + a \times 0 = a \times a.$$

Therefore

$$a \times 0 = 0,$$

because 0 is that number which when added to any number b gives b.

Now students are presumably prepared to appreciate the proof that a negative integer times a positive integer is negative. That is, $-3 \times 4 = -12$. We start with

$$(1) \qquad a \times [b + (-b)] = a \times 0 = 0$$

because $b + (-b) = 0$. However, by the distributive axiom

$$(2) \qquad a \times [b + (-b)] = a \times b + a \times (-b).$$

Hence the right sides of (1) and (2) are equal, so that

$$a \times b + a \times (-b) = 0.$$

Then $a \times (-b)$ must be $-(a \times b)$ because $a \times (-b)$ added to $a \times b$ gives 0 and this is true only for the additive inverse of $a \times b$, namely, $-(a \times b)$.

Finally, if in place of (1) we start with

$$-a \times [b + (-b)]$$

and carry through the same steps with $-a$ replacing a, we can prove that

$$-a \times -b = a \times b.$$

To introduce signed (positive and negative) numbers another method is often used. One starts with the numbers, 1,2,3,4, . . .; these are the counting numbers except for 0, and are called the natural numbers. Then an integer is defined as the equivalence class of ordered pairs of natural numbers. What this means is the following. An ordered pair of natural numbers is the pair (7,5). This, intuitively, means $7 - 5$. However, (6,4), (4,2), (8,6), and millions of other pairs represent the same integer. Two such ordered pairs (a,b) and (c,d) are called equivalent if $a + d = b + c$. Hence the integer 2 is the class of all ordered pairs equivalent to, say (7,5). The "merit" of this definition is that one can, using only the natural numbers, introduce the ordered pair (5,7), which intuitively represents $5 - 7$, or -2. Again (5,7), (4,6), (6,8), and so on are the same negative integer, -2. The integer which we usually denote by 0 is the class of ordered pairs (5,5), (6,6), (7,7), and so on. In the previous method we had to create the negative numbers. In the present method we "construct" the negative numbers.

Under this approach the operations with positive and negative integers are defined in terms of the ordered pairs. Thus, the sum of (7,5) and (6,3) is (13,8). Intuitively we have $2 + 3 = 5$, but the logical development calls for the former. More generally $(a,b) + (c,d) = (a + c, b + d)$. We see that for the negative integers the definition of sum works in the same way. Thus (5,7) + (3,6) = (8,13) or, intuitively, $-2 + -3 = -5$. The ordered pair definition can be used to introduce subtraction, multiplication and division of integers, and miraculously one obtains the usual laws for handling positive and negative integers.

To introduce fractions an approach similar to that used to introduce negative numbers is employed. The student is asked to find the value of x for which $3x = 6$. Clearly $x = (1/3) \times 6$ or $6/3$. He is then asked to find the x for which $3x = 7$. No integer will satisfy this equation. Hence we create new numbers, the fractions. In particular we create for x the number $7/3$, which means $(1/3) \times 7$, and we agree that $3 \times 1/3 = 1$. Having introduced the fractions we agree further that the commutative, associative and distributive axioms are to apply to them. We can then prove that the usual operations for the addition, subtraction, multiplication and division of fractions apply.

In the case of fractions, too, some texts introduce them as ordered pairs of signed integers. Thus (3,5) is a fraction which intuitively means $3/5$. The operations are defined in terms of the ordered pairs and then one can prove that addition and multiplication are commutative, associative and distributive.

The deductive approach encounters a serious obstacle, at least on the elementary level, in treating irrational numbers. The reader may recall that numbers such as $\sqrt{2}$, $\sqrt{5}$, $\sqrt[3]{4}$ and the like are also members of the number system. The full logical treatment of these numbers is much too difficult to be understood by tyros, and even if the students could assimilate the material the time required to teach it would be inordinate. Hence the texts usually compromise. They "motivate" the introduction of such numbers by first showing that to solve

$$x^2 = 4$$

we take the square root of both sides and obtain

$$x = \pm 2.$$

Similarly to solve

$$x^2 = 2$$

one takes the square root of both sides and obtains

$$x = \pm\sqrt{2}.$$

Thus one is led to introduce the irrational numbers. One can indeed prove, and many texts do, that $\sqrt{2}$ is not a rational number; that is, it is not equal to a quotient of two integers. So it is clear that objects such as $\sqrt{2}$ are new kinds of numbers. But the "motivation" used above does not suffice to introduce all irrational numbers, π, in particular. It also does not serve as a logical basis on which to build the properties of irrational numbers. Consequently these properties have to be postulated. For example, it is true, if a and b are greater than zero, that $\sqrt{a}\,\sqrt{b} = \sqrt{ab}$. But since it cannot be proved on an elementary level the texts just assert it and give it a name. It is called the product property of square roots. Likewise the fact that $\sqrt{a/b} = \sqrt{a}/\sqrt{b}$ cannot be proved, and so is merely asserted and labeled the quotient property for square roots.

Other difficulties arise in the introduction of complex numbers, that is, numbers of the form $3 + \sqrt{-1}$, $2\sqrt{-2}$, etc. The new curriculum, in this case like the traditional one, usually invents $\sqrt{-1}$ as the solution of $x^2 = -1$ and then forms the complex numbers. Alternatively it introduces complex numbers as ordered pairs of real numbers. It then defines the arithmetic operations with complex numbers and proves that the associative, commutative and distributive properties apply to these operations.

The logical foundation of arithmetic now serves to build algebra. Algebra is distinguished from arithmetic in that it deals with expressions involving letters such as $3x^2 + 5x + 2$ and operates with such expressions. Since the letters stand for numbers and all numbers obey the same laws, the letters obey these laws. Let us consider one algebraic proof. We shall prove that if $a \times b = 0$, then either a is 0 or b is 0 or both are zero.

If $a = 0$ the theorem is proved. If a does not equal zero then a property of the number system tells us that there is an inverse, $1/a$. Then, since $a \times b = 0$ and any number multiplied by 0 yields 0,

$$\frac{1}{a} \times (a \times b) = \frac{1}{a} \times 0 = 0.$$

By the associative axiom of multiplication

$$\left(\frac{1}{a} \times a\right) \times b = 0.$$

Then since $(1/a) \times a = 1$,

$$1 \times b = 0;$$

or since

$$1 \times b = b,$$
$$b = 0.$$

Thus the theorem is proved.

Some of the textbook authors wish to cut short the logical development of the number system or to "innovate." The authors of one widely used text, in introducing negative numbers, list many of the usual properties: the sum of any two integers is an integer, the commutative, associative, and distributive properties and such peculiar ones as the property of the opposite of a sum: $-(a + b) = -a + (-b)$. They give examples of the uses of these properties which students are asked to imitate. Some thirty or forty properties are finally listed and the students are expected to learn and apply them. The authors do not say whether these properties are axioms nor do they prove the assertions. The net effect is confusion about what is proved from axioms and what are rules. In effect the authors are handing down rules despite their claims to be teaching deductive mathematics.

Our examples of the way in which deduction is applied to arithmetic and algebra have been very simple. One can readily imagine how lengthy and involved the proofs are when doing more complicated arithmetic and algebra.

The deductive approach to Euclidean geometry is essentially the one most adults learned in high school. Hence it is not necessary to illustrate it here. However, we shall have more to say about proof in geometry in the next chapter.

Since the major innovation of the new mathematics is the deductive approach to traditional subject matter, let us try to determine what pedagogical merit it may have. In particular, does it impart understanding of mathematics?

A number of considerations oblige us to answer this question negatively. First, let us examine how mathematics itself developed and see whether this history furnishes any evidence favoring one conclusion or another. After all, mathematics was created by human beings who certainly understood the subject. How did the masters

Euclid, Archimedes, Newton, Euler, and Gauss come to understand mathematics?

Mathematics in a significant sense begins with the contributions of the Egyptians and Babylonians during the period of roughly 3000 to 300 B.C. These two peoples created the rudiments of arithmetic, algebra, and geometry. In arithmetic they worked with the positive whole numbers and fractions. Negative numbers were unknown to them and even the zero was not introduced despite the fact that Babylonians used positional notation in base sixty to write large numbers. That is, in their symbolism a number such as 125 meant $1 \cdot (60)^2 + 2 \cdot 60 + 5$, just as in our base ten 125 means $1 \cdot 10^2 + 2 \cdot 10 + 5$. Such a system of writing quantity almost cries out for a zero, because to write 105 one needs the 0 to indicate that the 1 is not in the "tens" position but in the "hundreds." Yet the Babylonians did not create the zero. Their numbers were ambiguous and one had to judge from the context as to what was intended.

In geometry all that the Egyptians and Babylonians could manage were formulas for perimeter, area and volume of simple geometrical figures. For any figure presenting difficulties, such as the perimeter and the area of a circle, the formulas were only approximately right. Thus for almost three thousand years, two civilizations rather highly developed in fields such as art, religion, commerce, astronomy, and architecture got no further in mathematics than the rudiments. Moreover, they accepted all their results on a purely empirical basis. The concept of deductive proof was never even entertained. Could these civilizations have profited from more extensive knowledge of mathematics? Undoubtedly. All we can conclude is that more sophisticated mathematics does not come easily to human minds. In fact, even the little the Egyptians and Babylonians achieved is unusual compared to what hundreds of other civilizations that had as much opportunity and need created in mathematics.

The first civilization in which mathematics can be said to have flourished is that of the classical Greeks; this civilization reached its zenith from 600 to 300 B.C. There is no question that the Greeks were of an unusual, even amazing, cast of mind. The classical Greek thinkers were indifferent to the needs of commerce, navigation, and practical matters generally, but they were intensely concerned about understanding the workings of nature. For this purpose they found geometry most suitable and it is in this area that they made their supreme contribution. The Greeks are also the people who first conceived of deductive mathematics. The goal was to obtain truths about nature and their plan was to start with some self-evident truths such as that two points determine a line and that all right angles are equal. Given these self-evident truths, or axioms, they planned to establish conclusions or theorems deductively. The theorems, then, would also be truths.

They did succeed in erecting several masterful structures, the foremost of which is Euclid's *Elements,* and this is the substance of the traditional high school geometry course. However, Euclidean geometry did not come into being in this deductive manner. It took three hundred years, the period from Thales to Euclid, of exploration, fumbling, vague and even incorrect arguments before the *Elements* could be organized. Thus the *Elements* is the finished and relatively sophisticated product of much cruder, intuitive thinking. Even this structure, intended to be strictly logical, rests heavily on intuitive arguments, pointless and even meaningless definitions and inadequate proofs, as the *nineteenth* century mathematicians realized. What is most relevant, however, is that this deductive system came about after the understanding of all that went into it was achieved. Moreover,

t is no accident that Euclidean geometry was the first
ubject to receive any extensive logical development; the
eason is that the intuition can be readily applied to in-
er geometrical facts and the very figures suggest methods
of proof.

It is also relevant that irrational numbers such as $\sqrt{2}$,
$\sqrt{3}, \sqrt[3]{5}$ and the like were not accepted as numbers dur-
ng the highest period of Greek culture. Why not? Be-
ause the whole numbers and fractions had an obvious
physical meaning whereas the irrational numbers did not.
The only intuitive meaning that one could attach to irra-
ionals was that they represented certain geometrical
engths, such as the diagonal of a square whose sides are
1. What then did the Greeks do? They rejected irra-
ionals as numbers and thought of them as lengths. In
fact they converted all of algebra into geometry in order
to work with lengths, areas and volumes that might otherwise have to be represented numerically by irrational
numbers and they even solved quadratic equations geo-
metrically.

The history of mathematics subsequent to the high
period of Greek culture is quite the opposite of what
one ordinarily conceives mathematics to be. The progress
that was made in the use of irrational numbers is due to
Alexandrian Greek civilization, which was a fusion of
the classical Greek, Egyptian and Babylonian civilizations, and to the Hindus and Arabs who were entirely
empirically oriented. It was the Hindus who decided that
$\sqrt{2} \sqrt{3} = \sqrt{6}$ and their argument was that these irrationals could be "reckoned with like integers," that is,
like $\sqrt{4} \sqrt{9} = \sqrt{36}$. Since the latter is obviously correct,
as can be seen by taking the square root of each number,
then so is $\sqrt{2} \sqrt{3} = \sqrt{6}$. Irrational numbers were gradually accepted because of their utility and because familiarity breeds uncriticalness.

Negative numbers, introduced by the practical-minded

Hindus about A.D. 600, did not gain acceptance for a thousand years. The reason: they lacked intuitive support. Some of the greatest mathematicians, Cardan, Vieta, Descartes, and Fermat, refused to work with negative numbers. The history of complex numbers is somewhat similar though they did not appear until about A.D. 1540 and only about two hundred years were required for these numbers to be used somewhat freely. A remark of the supreme mathematician Carl Friedrich Gauss is very pertinent. As is well known, he was one of the men who discovered the geometrical representation of complex numbers, and about this he said in 1831, "Here [in this representation] the demonstration of an intuitive meaning of $\sqrt{-1}$ is completely grounded and more is not needed in order to admit these quantities into the domain of the objects of arithmetic." Neither Descartes, Fermat, Newton, Leibniz, Euler, Lagrange, Gauss, or Cauchy could have given a definition of negative numbers or complex numbers, or irrationals for that matter. Yet all of them managed to work with these numbers quite satisfactorily, at least so far as their times employed these numbers. The history of the entire number system is pertinent not only because it shows how it was developed but also because algebra and analysis (the calculus and higher branches built on it) obviously utilize the number system, and whatever basis there was for the number system had to serve as the basis for algebra and analysis.

In the realm of algebra history has another significant tale to tell. The use of a letter to represent a fixed but unknown number dates from Greek times. However, the use of a letter or letters to stand for a whole class of numbers was not conceived of until the late sixteenth century. At that time François Vieta introduced expressions such as $ax + b$ where a and b can be any (real) number.* The great merit of such general expressions is that

* Actually Vieta did not admit negative values for a, b and x.

hatever we can prove about them is correct for all alues of a, b, and x. Thus if we learn to solve the quadratic equation $ax^2 + bx + c = 0$, we can solve all quadratic equations, because a, b, and c can stand for any umbers. The use of letters to stand for any numbers r even a restricted class of numbers is certainly, then, n enormous contribution and a seemingly simple one nce it is pointed out. Yet during all the centuries that he Babylonians, Egyptians, Alexandrian Greeks, Hindus, nd Arabs worked in algebra the idea of using letters or a class of numbers did not occur. These peoples did heir algebra by working with concrete expressions such s $3x^2 + 5x + 6 = 0$. That is, they always used numerical coefficients, and in fact most did not even use a symbol such as x for the unknown. They used words.

Why was the use of letters for general coefficients so ong delayed? The answer would seem to be that this device constitutes a higher level of abstraction in mathematics, a level farther removed from intuition. It is more difficult to think about $ax^2 + bx + c = 0$ than about $3x^2 + 5x + 6 = 0$. Yet to reason deductively about the algebraic procedures for significant general expressions became possible only after these general coefficients were introduced.

The history of the calculus is equally instructive. We shall not enter into the details of the concepts that lie at the foundation of that subject. But it may be sufficient to point out a few facts about its development. The great names in the creation of the calculus are, of course, Isaac Newton and Gottfried Wilhelm Leibniz. However, preceding them, Descartes, Fermat, Cavalieri, Pascal, Roberval, Barrow, and at least a dozen other well-known men made significant contributions. Despite the fact that so much was done before they did their work, neither Newton nor Leibniz could formulate correctly the basic concepts of the calculus. Newton wrote three major

papers on the calculus and he put forth three editions o
his masterpiece, *The Mathematical Principles of Natura
Philosophy*. In each of these he gave a different explanation of the basic concept now called the derivative. No
one of these would be accepted from a beginning studen
of the calculus today. Leibniz in many papers was equally
unsuccessful. His first paper was described by the famou
Bernoulli brothers, James and John, as "an enigma
rather than an explication."

There were many attacks on the work of both Newton
and Leibniz. Newton did not respond but Leibniz did.
He objected to "overprecise critics" and he argued that
we should not be led by excessive scrupulousness to reject the fruits of invention. However, the defects were
there and the attacks continued throughout the eighteenth century. The difficulties in clarifying the basic
concepts of the calculus were so great that the famous
eighteenth-century mathematician Jean LeRond d'Alem-
bert had to advise students, "Persist and faith will come
to you."

In view of the vague, unclear and even incorrect foundations of the calculus one might expect that the subject
would collapse. But before the adequate deductive structure was created not only had the calculus been extended
and successfully applied but the vast subjects of ordinary
and partial differential equations, the calculus of variations, differential geometry, and the theory of functions
of a complex variable had been erected on the calculus.
How did mathematicians achieve these tremendous creations? Clearly they thought intuitively. Physical arguments, pictures, generalizations based on simple known
cases, and experience with mathematics all helped them
to draw correct conclusions.

It is highly significant that the logical foundations of
the number system, algebra, and analysis (the calculus
and its extensions) were not erected until the last part of

the nineteenth century. In other words, during the centuries in which the major branches of mathematics were
built up there was no logical development for most of it.
Apparently the intuitions of great men are more powerful than their logic.

What can we infer from this history? It seems clear
that the concepts which have the most intuitive meaning,
the whole numbers, fractions, and geometrical concepts,
were accepted and utilized first. The less intuitive ones,
irrational numbers, negative numbers, complex numbers,
the use of letters for general coefficients, and concepts of
the calculus, required many centuries either to be created
or to be accepted. Moreover, when they were accepted it
was not the logic that induced mathematicians to adopt
them but arguments by analogy, the physical meaning of
some concepts, and the obtainment of correct scientific
results. In other words, it was intuitive evidence that induced mathematicians to accept them. The logic always
came long after the creations and evidently was harder to
come by. Thus the history of mathematics suggests,
though it does not prove, that the logical approach is
more difficult.

There are many who would base another argument
against the deductive approach on the evidence of history. The argument is essentially that each person must
go through about the same experiences that his fore-
fathers did if he is to attain the level of thought that many
generations have achieved. This argument has been advanced by many great mathematicians who have concerned themselves with pedagogy. Henri Poincaré, one
of the greatest mathematicians of recent times, said in
his *Foundations of Science* (p. 437), "The zoologists
maintain that in a brief period the development of the
embryo of an animal recapitulates the history of its an-
cestors of all geological epochs. It appears that it is the
same in the development of the mind. The task of the

educator is to make the mind of a child go through what his fathers have experienced, to pass rapidly through certain stages but not to omit any. For this purpose, the history of the science ought to be our guide."

The same thought has been expressed by one of the great mathematicians and teachers of the late nineteenth and early twentieth centuries. Felix Klein, in his *Elementary Mathematics from an Advanced Standpoint* (Dover reprint, 1945, Vol. 1, p. 268), says, "In concluding this discussion of the theory of assemblages [sets] we must again put the question which accompanies all of our lectures: *How much of this can one use in the schools?* From the standpoint of mathematical pedagogy, we must of course protest against putting such abstract and difficult things before the pupils too early. In order to give precise expression to my own view on this point, I should like to bring forward the biogenetic fundamental law, according to which the individual in his development goes through, in an abridged series, all the stages in the development of the species. Such thoughts have become today part and parcel of the general culture of everybody. Now, I think that instruction in mathematics, as well as in everything else, should follow this law, at least in general. Taking into account the native ability of youth, instruction should guide it slowly to higher ideas, and finally to abstract formulations, and in doing this it should follow the same road along which the human race has striven from its naive original state to higher forms of knowledge. It is necessary to formulate this principle frequently, for there are always people who, after the fashion of the medieval scholastics, begin their instruction with the most general ideas, defending this method as the 'only scientific one'. And yet this justification is based on anything but truth. To instruct scientifically can only mean to induce the person to think scientifically, but by no means to confront him, from the beginning, with cold, scientifically polished systematics.

"An essential obstacle to the spreading of such a natural and truly scientific method of instruction is the lack of historical knowledge which so often makes itself felt. In order to combat this, I have made a point of introducing historical remarks into my presentation. By doing this I trust I have made it clear to you how slowly all mathematical ideas have come into being, how they have nearly always appeared first in rather precursory form, and only after long development have crystallized into the definitive form so familiar in systematic presentation."

There is not much doubt that the difficulties the great mathematicians encountered are precisely the stumbling blocks that students experience and that no attempt to smother these difficulties with logical verbiage can succeed. If it took mathematicians a thousand years from the time first-class mathematics appeared to arrive at the concept of negative numbers—and it did—and if it took another thousand years for mathematicians to accept negative numbers—as it did—we may be sure that students will have difficulties with negative numbers. Moreover, the students will have to master these difficulties in about the same way that the mathematicians did, by gradually accustoming themselves to the new concepts, by working with them and by taking advantage of all the intuitive support that the teacher can muster.

One could of course argue that the growth of mathematics may indeed have proceeded as just described, but now that we have the proper logical structures for the number system, the calculus and other branches, we need not ask the students to repeat the fumblings of the masters. We can give students the correct approaches and they will understand them. This argument can be countered by the fact that the greatest mathematicians did try to build the logical foundations for the various

subjects but failed for centuries to do so. Their failure
should serve as some evidence that the logical approaches
are not easy to grasp. One can compress history and
avoid many of the wasted efforts and pitfalls, but one
cannot eliminate it. Of course, our students may be superior to the best mathematicians of the past.

But we shall not insist on the evidence of history.
There are other weighty arguments against a purely
deductive approach to elementary mathematics.

We might note, first of all, that we are not without experience in presenting mathematics deductively. Euclidean geometry has been presented in this manner for
several centuries. Moreover, the intuitive meaning of
this geometry is also evident to the student. Yet students
have not been more successful in mastering geometry
than algebra; nor have they left the geometry course with
a feeling of elation because they have finally understood
a branch of mathematics. If, then, the evidence of history is not convincing, there must be other pedagogical
arguments against a logical approach. They are not hard
to find.

By the middle of the nineteenth century the various
types of numbers and their properties were established
on the basis of the uses made of them. Likewise, the
properties of functions, derivatives and integrals used
in the calculus were accepted on the basis of what seemed
evident for the simplest functions or on the basis of the
physical truth of the results obtained. The mathematicians then set about constructing logical foundations for
the properties they had employed. In fact, the logic had
to *justify* those properties, rather than determine them.
Hence a very artificial and complicated structure of axioms and theorems was erected. The purpose of this structure was to satisfy the needs of professional mathematicians who insist on deductive structure, but it was never
intended as a pedagogical approach. Yet it is these logi-

cal foundations that the new mathematics employs to cultivate understanding.

The fact that utility determines the logical approach rather than the other way around is so basic in mathematics that the point warrants emphasis. Let us consider an example. To use addition of fractions in most real situations, say, to add $1/2$ and $1/3$, we change both to sixths and then add $3/6$ and $2/6$ to obtain $5/6$. However, when we multiply fractions we multiply the numerators and multiply the denominators so that $1/2 \times 1/3 = 1/6$. We could "add" fractions by adding the numerators and adding the denominators and obtain $1/2 + 1/3 = 2/5$. Why don't we use this latter method? It is simpler. But it does not fit experience. Having adopted a useful definition of addition the logical properties of addition must follow from the definition.

As another example we could consider matrix multiplication. It so happens that the uses to which matrices are put requires that the multiplication be noncommutative, though we could define a multiplication which is commutative. Since the multiplication must be noncommutative, the logical foundations of the theory of matrices must be suited to this fact. Therefore, logic does not dictate the contents of mathematics; the uses determine the logical structure. The logical organization is an afterthought and in a real sense is gilt on the lily.*

In fact, if a student is really bright and he is told to cite the commutative axiom to justify, say, $3 \times 4 = 4 \times 3$, he may very well ask, Why is the commutative axiom correct? The true answer is, of course, that we accept the commutative axiom because our experience with groups of objects tells us that $3 \times 4 = 4 \times 3$. In other words, the commutative axiom is correct because $4 \times 3 = 3 \times 4$ and not the other way around. The normal student will parrot the words "commutative axiom"

* See Chapter 8, p. 105 for the definition of a matrix.

and he will, as Pascal put it in his *Provincial Letters,* "fix this term in his memory because it means nothing to his intelligence."

To repeat, the uses to which mathematics is put must be known to construct the logical foundations. Poincaré noted that in building up the number system from the positive whole numbers there are many different constructions one can make. Why do we take one rather than another? "The choice is guided by the recollection of the intuitive notion in which this construction took place; without this recollection, the choice appears unjustified. But to understand a theory it is not sufficient to show that the path that one follows does not present obstacles; it is necessary to take account of the reasons that one chooses that path. Can one ever understand a theory if one builds it up right from the start in the definitive form that rigorous logic imposes, without some indications of the attempts which led to it? No; one does not really understand it; one cannot even retain it or one retains it only by learning it by heart."

Poincaré's point may be made by considering the game of chess. To play the game it is not sufficient to know the rules for moving the pieces. With just this much knowledge one can see that a correct move has been made but one will not understand why one move has been made rather than another. The inner reason or strategy is not apparent. In the case of mathematics the inner reason is usage.

Actually the logical approach is misleading. In extending the number system from the natural numbers to the various other types, the new curriculum insists that the commutative and associative properties of the operations be retained. Why do we insist on these properties? The teachers know that the uses of the numbers call for these properties, but the students get the impression that these are necessary properties of all mathematical quantities.

Why, then, do we not extend the commutative property to the multiplication of matrices? The answer is that the uses to which matrices are put require a noncommutative multiplication. The logical approach gives the student an entirely false impression of how mathematics develops.

Most proofs presented to students are also artificial for additional reasons. When a mathematician seeks to prove a theorem he suspects is correct, he uses any means, however clumsy, indirect or devious, but perhaps more natural to the creative process, to make the proof. Once the theorem is proven he or his successors, now able to see how the essential difficulty was overcome, can usually devise a smoother or more direct proof. Some theorems have been reproven several times, each successive proof remodeling and simplifying the previous one and often including generalizations or stronger results. Thus, the final theorem and proof are far from the original natural thoughts. One should expect, then, that the student facing such a reworked, polished, and possibly more complicated result would not be able to grasp it. Beyond attempting to refashion a proof to secure brevity and perhaps elegance, mathematicians like to erect logical structures in which many theorems follow from a small number of axioms. Fitting a theorem into such a structure may call for further juggling of the proof and cause still more difficulty to the student who seeks to understand it.

Many teachers, having delivered a series of such theorems and proofs, walk out of their classrooms very satisfied with themselves. But the students are not satisfied. They have not understood what was going on and all they can do is memorize what they heard. They were not involved in the original thinking and derived no insight from the polished proofs. The deductive approach has been likened to the action of a fox who effaces his footsteps with his tail.

The insistence on a logical approach also deceives the student in another way. He is led to believe that mathematics is created by geniuses who start with axioms and reason directly from the axioms to the theorems. The student, unable to function in this manner, feels humbled and baffled but the obliging teacher is fully prepared to demonstrate genius in action. Perhaps most of us do not need to be told how mathematics is created, but it may help to listen to the words of Felix Klein. "You can often hear from non-mathematicians, especially from philosophers, that mathematics consists exclusively in drawing conclusions from clearly stated premises; and that in this process, it makes no difference what these premises signify, whether they are true or false, provided only that they do not contradict one another. But a person who has done productive mathematical work will talk quite differently. In fact these people [the non-mathematicians] are thinking only of the crystallized form into which finished mathematical theories are finally cast. However, the investigator himself, in mathematics as in every other science, does not work in this rigorous deductive fashion. On the contrary, he makes essential use of his imagination and proceeds inductively aided by heuristic expedients. One can give numerous examples of mathematicians who have discovered theorems of the greatest importance which they were unable to prove. Should one then refuse to recognize this as a great accomplishment and in deference to the above definition insist that this is not mathematics? After all it is an arbitrary thing how the word is to be used, but no judgment of value can deny that the inductive work of the person who first announces the theorem is at least as valuable as the deductive work of the one who proves it. For both are equally necessary and the discovery is the presupposition of the later conclusion." It is intellectually dishonest to teach the deductive approach as though the results were acquired by logic.

The logical approach produces practical complications. If a student has to show, for example, that $4ab(ab + 3ac) = 4a^2b^2 + 12a^2bc$ and if he has to justify each step, he will have to think carefully and give reasons for so many steps that he will take minutes to do what he should do almost automatically on the basis of experience with numbers. It is far preferable that the student should become so familiar with the basic properties such as distributivity, commutativity and associativity that he does not realize he is using them.

Asking the students to cite axioms in the elementary operations with numbers is like asking an adult to justify each action he takes after getting up in the morning. Why should he bathe? Why should he brush his teeth? Why wear clothes? And so on. If a man should consciously consider and answer these questions he would never get to work. Most of what he does in the morning must be habitual.

There is a story that a centipede was walking along leisurely when it met a toad. The toad remarked to the centipede, "Isn't it wonderful? You have one hundred feet and yet you know when to use each one." Thereupon the centipede began to think about which foot to use next and was unable to move.

Students should learn to become habitual about the elementary operations with numbers so that they do not have to think about them. After seeing that four sets of three blocks and three sets of four blocks both amount to twelve blocks, the students should accept the commutative principle as so obvious that it need not be mentioned. Moreover, instead of being obliged to consider this property for the natural numbers, the signed numbers, the rational numbers, and so on, he should be induced to believe that all numbers possess these properties. In fact, we should do all we can to make the elementary operations so habitual that students do not have to think about them any more than one thinks when he ties his shoelaces. We should be grateful that students will accept unquestioningly facts that seem entirely reasonable to them whether on the basis of experience with numbers or intuitive arguments. If students do not see readily that $3 \times a = a \times 3$, it is not because they lack familiarity with the commutative principle but rather because they fail to understand that a is just a number. When the time to teach a noncommutative operation arrives, then the concept of commutativity can be discussed.

The need to make some of the work automatic was stressed by a man who certainly understood the role of deductive proof. The philosopher Alfred North Whitehead said in *An Introduction to Mathematics,* "It is a profoundly erroneous truism, repeated by all copybooks, and by eminent people when they are making speeches, that we should cultivate the habit of thinking of what we are doing. The precise opposite is the case. Civilization advances by extending the number of important operations which we can perform without thinking about them. Operations of thought are like cavalry charges in a battle—they are strictly limited in number, they require fresh horses, and must only be made at decisive moments."

In many areas the present emphasis on the logical approach is sheer hypocrisy. What mathematician used the logical development of the complex number system to justify his operations with real or complex numbers? Yet this is what is taught to students as the way to learn the "truth" about numbers. How many mathematicians have ever satisfied themselves that $\sqrt{2}^{\sqrt{3}}$ is defined in the theory of irrational numbers or have examined the rigorous proof that $\sqrt{2}\,\sqrt{3} = \sqrt{3}\,\sqrt{2}$? How many have ever worked through a rigorous development of Euclidean geometry, which would call for no reliance upon figures? Felix Klein did not hesitate to admit, "To follow a geometrical argument purely logically without having the figure on which the argument bears constantly before me is for me impossible."

As a matter of fact the attempt to be completely deductive ensnares the teacher in a trap. It is often necessary to include a proof which even the logic-oriented teachers concede to be too difficult for the student, such as the proof of the formula for the area of a circle in plane geometry. Many texts evade the issue by adopting as an axiom that the area $=\pi r^2$, where r is the radius. Surely if one can adopt axioms at will there is no need to prove anything. The only lesson the student will learn from such presentations is that if he is stuck he can adopt an axiom. Whereas in other matters students are asked to digest trivia thoroughly, the area axiom asks them to swallow a camel whole.

Moreover, by adopting axioms freely numerous texts employ as many as seventy or eighty axioms. Since students are required to make proofs by citing axioms where these are the justifications for steps, the students are obliged to remember the seventy or eighty axioms. This is an intolerable burden, an impossible one for students to carry. Yet such books claim to avoid memorization and to teach thinking and understanding. The mathematics teacher can no more afford to be profligate with axioms than to be parsimonious.

Other measures to avoid difficult proofs are equally harmful. In the presentation of the real number system the high school texts proceed axiomatically from the natural numbers. But when they get to the irrational numbers, whose logical development the authors recognize to be too difficult for the student, they resort to the number line. They show how integers and fractions can be attached to points on the line and then note that some points do not have numbers assigned to them. The irrational numbers are then introduced as the numbers to be attached to these points. If the logical presentation of the rational numbers had any value it is dissipated by this meaningless introduction of the irrationals.

Every curriculum claims to teach the student how to discover results for himself. Discovery, contrasted with and opposed to the passive and uncritical acceptance of finished or polished statements of theorems and proofs, amounts to the creation or re-creation of mathematics by the student, possibly with the guidance of the teacher. How do the mathematicians create? Their first task is to divine a possible theorem. Having gotten this far they must do further creative work to find a proof. As the supreme mathematician Carl Friedrich Gauss put it, "I have got my result but I do not know yet how to get [prove] it." In the creative work imagination, intuition, experimentation, judicious guessing, trial and error, the use of analogies even of the vaguest sort, blundering and fumbling enter. Deductive proof plays little if any role.

Creativity presupposes flexibility in solving problems, and any ideas from any domain of mathematics should be entertained whether or not they fall within the confines of a particular axiomatic structure. The latter, in fact, acts as a straitjacket on the mind.

What does logic contribute to the creation of concepts? What deductive process tells us to consider similar triangles or to consider the altitudes or the medians of a triangle? Logic alone is as incapable of leading to new ideas as grammar alone is incapable of leading to poetry and the theory of harmony to music. Creation has sometimes been described as the following process. The mathematician says *A*, writes *B*, means *C*, but *D* is what it should be. And *D* is in fact a splendid idea which emerges from tidying up the mess.

Some of the greatest ideas in mathematics are not at all a matter of logic. Perhaps the best example is the realization that non-Euclidean geometry is applicable to physical space. The logical side, namely, pursuing the consequences of assuming a non-Euclidean parallel axiom, was a relatively simple task and was performed by Saccheri, Lambert, Legendre, Schweikart, Taurinus, and many others. But it was Gauss who first recognized that these new geometries are as applicable as Euclidean geometry. The consequences for mathematics were as revolutionary as the very creation of mathematics itself.

A foremost mathematician of our times, Henri Lebesgue, pointed out the subordinate role of logic. "No discovery has been made in mathematics, or anywhere else for that matter, by an effort of deductive logic; it results from the work of creative imagination which builds what seems to be truth, guided sometimes by analogies, sometimes by an esthetic ideal, but which does not hold at all on solid logical bases. Once a discovery is made, logic intervenes to act as a control; it is logic that ultimately decides whether the discovery is really true or is illusory; its role therefore, though considerable, is only secondary."

Thus the concentration on the deductive approach omits the vital work. It destroys the life and spirit of mathematics. The deductive formulation does dress up the real activity, but it conceals the flesh and blood. It is like the clothes which make the woman but are not the woman. It is the last act in the development of a branch of mathematics and, as one wise professor put it, when this is performed the subject is ready for burial. Logic may be a standard and an obligation of mathematics but it is not the essence any more than the grammatical structure is the essence of the Bible or Shakespeare's works. The deductive structures which many mathematicians like to call mathematics are the dried-up stalks of the living plants. They are empty formalisms as opposed to real contents; they are shells of palaces.

Another argument advanced by the advocates of the new mathematics is that their emphasis on logical structure teaches students to think deductively. This is probably correct. But even if they do teach deductive logic, why is this so important? It is not the kind of thinking that is useful in everyday life. The big problems and even the little ones that human beings are called upon to solve in life cannot be solved deductively. There are no self-evident axioms from which one can deduce what career to follow, whom to marry, or even whether to go to the movies. The real decisions call for judgment, and this is entirely different from deductive reasoning which leaves no room for judgment. The legal mind, the business mind, and the political mind are much more relevant.

The modernists often espouse another argument for their approach. Mathematics can be enjoyed as a game played according to certain rules. But the purely axiomatic approach to mathematics must surely raise the question in intelligent students' minds, "Why should I play this game according to these particular rules?"

In view of the many pedagogical shortcomings in the logical approach to mathematics it is not surprising that many perceptive mathematicians (there are some nonperceptive ones) have spoken out against the logical approach. Descartes deprecated logic in rather severe language. "I found that, as for Logic, its syllogisms and the majority of its other precepts are useful rather in the communication of what we already know or . . . in speaking without judgment about things of which one is ignorant, than in the investigation of the unknown." Roger Bacon said, "Argument concludes a question but it does not make us feel certain, or acquiesce in the contemplation of a truth, except the truth also be found to be so by experience." Pascal pointed out, "Reason is the slow and tortuous method by which those who do not understand the truth discover it."

The logical approach is reminiscent of the reply Samuel Johnson gave to a man who persisted in asking for an explanation of his remarks. Johnson said, "I have found you an argument but I am not obliged to find you an understanding." The sterile, desiccated axiomatic approach has not promoted understanding. The formal logical style is one of the most devitalizing influences in the teaching of school mathematics. An ordered logical presentation of mathematics may have aesthetic appeal to the mathematician but serves as an anaesthetic to the student.

These arguments against an exclusively deductive approach to mathematics are not meant to imply that one should reject completely the use of deductive proof. It has a place which we shall discuss later (Chapter 11), but the most important fact about deductive proof is to keep it in its place.

Perhaps, after all, there is some merit to the logical approach to mathematics. As Bertrand Russell put it in his *The Principles of Mathematics* (p. 360), "It is one of the chief merits of proofs that they instill a certain scepticism as to the result proved." Henri Lebesgue pointed out another value of deductive proof: "Logic makes us reject certain arguments but it cannot make us believe any argument." One must respect but suspect mathematical proofs. Since one of the main objectives of mathematics education is to instill scepticism in the student, he is deriving at least one benefit from the current logical extravaganzas.

5

Rigor

"Geometry is nothing if it be not rigorous, . . .
The methods of Euclid are, by almost universal
consent, unexceptionable in point of rigour."
                    Henry John Stephen Smith (1873)

The modern mathematics leaders are not content with
a deductive approach to mathematics. They wish to present a rigorous deductive development.

Let us first be clear as to the distinction between a
deductive development and a rigorous deductive development. Euclidean geometry as most adults learned it in
high school, which is essentially Euclid's own presentation in his *Elements* of about 300 B.C., is deductive.
However, it is not rigorous. The distinction lies in the
fact that Euclid and his successors up to recent times
used implicitly axioms and theorems that are so obviously true that they either did not realize they were using
them (just as we are usually unaware of breathing air)
or they thought there was no need to assert them or
mention them in proofs. Thus, it is obvious that a line
divides a plane into two parts and that a triangle has an

nside and an outside. If we have three points on a line
t seems abundantly clear that one and only one of these
three points lies between the other two.

Or consider another example. Suppose two circles are
drawn with the distance between the two centers less
than the sum of the two radii (Fig. 5.1). Euclid did not
hesitate to affirm that the two circles intersect in two
points. This fact is not guaranteed by Euclid's axioms.
One could argue that the structure of the circle might be
such that just at the two places where the circles are pre-
sumed to intersect either circle or both may fail to have
points on them. The argument is technically correct.
However, Euclid's concept of the circle, like our intuitive
concept, was what one might describe as a continuous
structure, that is, one with points everywhere along the
circumference.

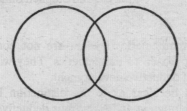

FIGURE 5.1

As we have stated, the above described facts and
others were used by Euclid and the traditional texts without mentioning them explicitly. The modernists believe
that students are disturbed by the use of unmentioned
assumptions and theorems and that, because the proofs
are not strictly complete, the students are hindered in
their understanding. It is also deceitful, say the modernists, to present proofs that are purportedly complete but
are not actually so.

The remedy for these "defects" in traditional Eu-

clidean geometry is to supply additional axioms and then
to prove every assertion, no matter how obvious, by deductive reasoning. The new axioms are of several types,
axioms of existence, axioms of order, and a continuity
axiom. Among the axioms that must be added are the
assertions that the line joining any two points is unique,
that the distance between two points is unique, that a
plane contains at least three points, and that a line divides the plane into two parts. There is also a continuity
axiom which in effect guarantees that the two circles previously described meet in two points and that if A and B
(Fig. 5.2) are on opposite sides of a line l, the line joining A and B actually intersects l in a point common to
the two lines. The wording of these additional axioms
varies somewhat from one modern text to another.

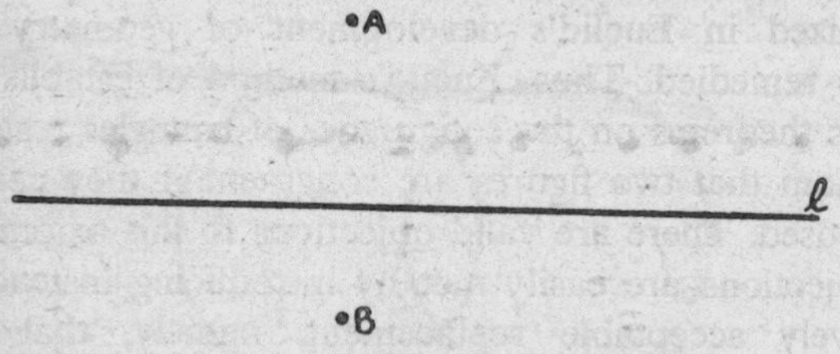

●A

━━━━━━━━━━━━━━━━━━━━━━━━━━━━━ l

●B

FIGURE 5.2

All the axioms and theorems required for a rigorous
deductive approach to geometry are spelled out in modern mathematics treatments. This involves many additional axioms and dozens of theorems in which the results are intuitively obvious. For example, one proves
that a line segment not only has a midpoint but a unique
midpoint, and that a triangle has an inside and an outside.

In algebra, too, rigor is incorporated. Thus, one does
not presuppose that the result of adding 4 and 6 is a
unique number. Conceivably there could be two answers.

One adopts what is called a closure axiom which asserts the existence and uniqueness of the sum. In dealing with the relation of equality one does not take for granted its obvious properties. One adopts as an axiom that $a = a$ and this property is called reflexiveness. Can one be certain that if $a = b$ then $b = a$? An axiom assures us that it is so. This property is called symmetry. Finally, one postulates that if $a = b$ and $b = c$, then $a = c$, and thus the transitivity of equality is assured.

It is true in algebra as in geometry that the addition of the many axioms which rigor demands requires that a number of obvious theorems be proved, even though students would otherwise use these theorems without recognizing that they had called upon them.

Is the incorporation of rigor a contribution to pedagogy? Let us note first of all that some of the defects now recognized in Euclid's development of geometry are readily remedied. Thus, Euclid's method of establishing various theorems on the congruence of triangles rests on his axiom that two figures are congruent if they can be superposed. There are valid objections to this axiom but the objections are easily met by introducing instead an intuitively acceptable replacement, namely, that two triangles are congruent if two sides and the included angle of one triangle are equal to two sides and the included angle of the other. In this instance the improvement on Euclid does not involve subtleties that are beyond the understanding of young people; hence, there is no reason not to make the change. However, there are serious objections to supplying most of the axioms and theorems that a truly rigorous deductive development requires.

Until about one hundred years ago mathematicians regarded Euclid's presentation as the model of rigor. It is true that occasionally a mathematician would either criticize the wording of an axiom or point out the need for an additional axiom. But no one took these criticisms seriously because they called for no more than minor changes or additions. It was the creation of non-Euclidean geometry in the first third of the nineteenth century which forced mathematicians to be more critical of what they had accepted in Euclidean geometry, and therewith they became aware that Euclid had used many axioms and theorems which are intuitively so obvious that he was not aware he was using them.

To ask students to recognize the need for these missing axioms and theorems is to ask for a critical attitude and maturity of mind that is entirely beyond young people. If the best mathematicians did not recognize the need for these axioms and theorems for over two thousand years how can we expect young people to see the need for them? That for two thousand years Euclidean geometry, as formulated by the presumably careless or naive Euclid, was regarded by the best mathematicians as the paradigm of rigor bears no weight with the advocates of rigorous axiomatics. Today's students, we are apparently supposed to believe, are sharper and are not satisfied with the cruder version presented by Euclid. Presumably, this too is a reason that they do not do well in geometry.

In presenting rigorous developments of the number system and of geometry we first have to make sure that students appreciate the gaps that rigor meets and then teach the rigor. Thus, in the case of geometry students use figures which automatically take care of such details as the order of points on a line, but which the rigorous presentation covers through suitable axioms. Hence, the teacher has to spend a great deal of time in making a student realize that he has accepted many facts on an intuitive or visual basis. Even if sufficient emphasis and persistence on the part of the teacher make students see the need for the missing axioms and theorems, they will draw a conclusion that will hardly endear mathematics o them. In a rigorous development of geometry and algebra many intuitively obvious theorems must be proved. The students' conclusion will be that mathematics is largely concerned with proving the obvious.

Further, the axioms that mathematicians adopt in a rigorous approach, though simple in the sense that they deal with basic properties of points, lines, and planes, are not obvious properties that one would naturally adopt. For example, one of the order axioms specifies that, given any three points on a line, one and only one lies between the other two. Another specifies that a line which cuts one side of a triangle necessarily cuts another side. Still another specifies that if the points of a line are divided into two sets such that all the points of one set precede, in the order of the points on a line, all those of the second set, then there is one and only one point which separates the points of the two sets. Such axioms are introduced to prove the simplest theorems of Euclidean geometry. In fact, many of the theorems are more obvious than the axioms used to establish them. Hence, the less obvious is used to prove the more obvious. But as far as the student is concerned the whole point of proof is just the reverse. Students will question what is being accomplished and perhaps even wonder whether teachers are sane.

Henri Poincaré, in an article on logic and intuition, struck at this folly. "When a student commences seriously to study mathematics, he believes he knows what a fraction is, what continuity is, and what the area of a curved surface is; he considers as evident, for example, that a continuous function cannot change its sign without vanishing. If, without any preparation, you say to him: No, that is not at all evident; I must demonstrate it to you; and if the demonstration rests on premises which do not appear to him more evident than the conclusion, what would this unfortunate student think? He will think that the science of mathematics is only ar arbitrary accumulation of useless subtleties; either he will be disgusted with it or he will amuse himself with it as a game and arrive at a state of mind analogous to that of the Greek sophists."

Almost three hundred years earlier Blaise Pascal said in his *Pensées*, "Never undertake to prove things that are so evident in themselves that one has nothing clearer by which to prove them."

Another consequence of incorporating all of the axioms a rigorous approach to Euclidean geometry requires is that a host of trivial theorems must be proved before the significant ones are reached. The number of minor theorems is so large that the major features of the subject fail to stand out. Moreover, the time consumed in proving the obvious theorems deprives the students of the time to study significant, deep theorems that are essential to progress in mathematics.

The rigorous development of a branch of mathematics is often so artificial that it is meaningless. No example is more pertinent than the logical development of the real number system. There were good reasons to axiomatize the number system, but the introduction of fractions and negative numbers as couples (see Chapter 4) with special definitions of the operations with these couples, clever as it may be, is so artificial, trumped-up and foreign to the intuitive meaning and uses of these numbers as to preclude understanding.

Many teachers might retort that the student has already learned the intuitive facts about the number system and is now ready for the appreciation of the rigorous version, which exemplifies mathematics. If the student really understands the number system intuitively, the logical development will not only *not* enhance his understanding but will destroy it. As an example of mathematical rigor no poorer choice could be made because the construction is so contrived. The development is so full of details and so stilted that it not only stultifies the mind but obscures the real ideas. Yet just this topic has now become the chief one in high school and college mathematics courses.

Some of the axioms used in a rigorous approach to the real number system must strike the students as absurd. They are asked to accept the closure axiom. In the case of the integers, for example, this reads: The sum of two integers is an integer. Were the students not properly forewarned by this axiom would they have thought that the sum of two integers is a cow? In the case of the rational numbers (positive and negative whole numbers and fractions) the uniqueness of inverses to addition and multiplication is stressed. Would the students have expected two answers for subtraction or two for division? These closure axioms may serve one purpose: they close the gates of the students' minds.

The distinction between deductive proof and rigor involves a further complication. Rigorous proof is not static. The demands of rigor are constantly changing and increasing in complexity. Mathematics grows like a tree. As the trunk, branches, and leaves increase, the roots go deeper. It is safe to say that no proof given at least up to 1800 in any area of mathematics, except possibly in the theory of numbers, would be regarded as satisfactory by the standards of 1900. The standards of 1900 are not acceptable today. Hence, the pedagogical significance of the constantly increasing standards of rigor is that if the texts continue to keep up with them the students will constantly be burrowing further down to the roots and will never get to see the tree proper. However, it is not necessary to pursue this rigor. Young people will not see the need for it any more than did the great mathematicians of a hundred years ago. What was intuitively acceptable two thousand years ago is still intuitively ac-

ceptable today. Moreover, the students can be much
more readily attracted to the fruits rather than to the
roots of mathematics.

It is rather ironic that the reformers, in their efforts to
be modern and up-to-date, decided to emphasize the
rigor of 1900. They are at least seventy-five years too
late. As the finishing touches were being put on the rigorous approaches that appeared satisfactory at the end
of the nineteenth century, difficulties in the logic of
mathematics were uncovered in just that branch of mathematics, set theory, which is the most heavily emphasized topic of the new mathematics. The difficulties,
which are euphemistically called the paradoxes of set
theory, but which are more accurately described as con-
tradictions in set theory, have not been resolved to the
satisfaction of all mathematicians, and the logic of mathematics has never been in a sorrier state. Indeed, there
is no agreement at all today on what constitutes a correct
mathematical proof, and it is fairly certain that an axiomatic approach to any branch of mathematics cannot be
adequate.

We cannot enter here upon the history of rigor or
upon the difficulties in building mathematics rigorously,
but we can state that many mathematicians are sufficiently skeptical of whether we shall ever attain rigor to
make sarcastic remarks such as "Logic is the art of going
wrong with confidence"; . . . "The virtue of a logical
proof is not that it compels belief but that it suggest
doubts . . ."; "A mathematical proof tells us where
to concentrate our doubts." The whole attempt to inject
rigor in mathematics has amounted to picking up jewels
only to discover serpents underneath.

Like deductive reasoning, rigor does play a role in
mathematics, but it is the concern only of professional
mathematicians who wish to ensure that the deductive
structures are sound. Such men, who have developed a

critical spirit after years of specialization, can see the need
for rigor and appreciate what it supplies. Without that
experience, the detailed and sophisticated axiomatics appear to be senseless, futile inventions. Hence, to offer it
to young students is to bewilder and baffle rather than to
aid them. In fact, the rigorous deductive presentations
introduced by the great mathematicians of the late nineteenth century and the early part of this century were
never intended as an aid to pedagogy. The great mathematicians who have taken an interest in pedagogy always
stressed that strict logical presentations are entirely subordinate to the substance, which is learned intuitively.
Rigor may save mathematics but it will surely lose the
pupils.

In view of the detriments to pedagogy which rigorous presentations impose, one might well ask why the
curriculum-makers incorporated it. We have pointed out
that the many curricula were written by groups of mathematicians and teachers drawn from all levels of the
mathematical world. Some of these writers, only recently
informed of the rigor of 1900, became enthusiastic about
presenting what they thought was the new face of mathematics. Of course, they confused what is logically prior
with what is pedagogically desirable. Others seeking novelty fastened on rigor. Among the many writers were
shallow, relatively ignorant mathematicians who took the
simple topics of elementary mathematics and made them
appear profound by cloaking them in what for young
people can be described only as prissy pedantry. They
sought thereby to give the impression of deep mathematical insight. It is easier to incorporate sophistication
in trivial matters than to give clear intuitive presentations
of the more difficult ideas. Certainly much of the rigor
in modern texts was inserted by limited men who sought
to conceal their own shallowness by a facade of profundity and by pedants who masked their pedantry under

the guise of rigor. One can rightly accuse them of pseudo-sophistication. If mathematical education of the traditional type has suffered from the martinets who imposed rote learning, the newer education will suffer more horribly from the rigor-mongers.

6

The Language of Mathematics

If you're anxious for to shine in the high aesthetic line as
  a man of culture rare,
You must get up all the germs of the transcendental
  terms, and plant them everywhere.
You must lie upon the daisies and discourse in novel
  phrases of your complicated state of mind,
The meaning doesn't matter if it's only idle chatter of a
  transcendental kind.
> And every one will say
> As you walk your mystic way,
"If this young man expresses himself in terms too deep
  for *me*,
Why, what a very singularly deep young man this deep
  young man must be!"

Sir W. S. Gilbert

One of the defects of the traditional curriculum, according to the modern mathematics leaders, is its imprecise
language. The looseness and ambiguities are supposedly

so numerous and so deplorable that students are seriously handicapped. The new curriculum claims to eradicate these defects by introducing precise language. Let us see how serious the defects have been and how they are purportedly eliminated.

To illustrate the inaccuracy of the traditional language the modernists give the following illustrations. "Peter has four balloons and Joe has five balloons. How many balloons do both have?" Almost everyone would understand the language to mean, "What is the sum of the number of balloons Peter has and the number Joe has?" and would answer nine. Not so, say the modernists. Both boys do not have any balloons, and they mean of course that they have no balloons in common.

A second illustration states, "Mary spent twelve cents for two pencils"; then it asks, "How much did she spend for each?" Most people would answer six cents because they would assume, in the absence of further information, that the two pencils are alike. The modernists object. There was no explicit statement to the effect that the pencils were alike.

Clearly these questions, if indeed imprecisely worded, could certainly be restated without requiring any special language. But the modernists believe that a drastic improvement in the language of mathematics is called for.

To secure precision they make the distinction between number and numeral. The symbol 7 is not a number but a symbol for a number. Other symbols for the same number are $3 + 4$, $5 + 2$, $8 - 1$, and many more. Students are expected to learn that they deal with numerals rather than numbers. To point up the necessity for the distinction the modern texts give the following example. One can say that the number 343 contains three digits. But $343 = 7^3$. Hence, one should be able to say that 7^3, which is the same number, contains three digits. The original statement should have been that the *numeral* 343 contains three digits.

Apropos of the distinction between number and numeral there is a story that one writing group entitled a chapter "Learning to Read and Write Large Numbers." When the group was apprised that one reads and writes numerals rather than numbers, they changed the word "numbers" to "numerals." But this title, they soon realized, could mean writing large-sized numerals and so the title was further changed to "Learning to Read and Write Numerals for Large Numbers." By this time no one understood what was meant.

Precision of language is further "insured" by using the language of sets. Sets are no more than collections of objects, for example, the set of all apples, the set of whole numbers, and the set of men. By using the concept of a set one can rephrase and presumably make precise many mathematical statements. Thus, instead of asking for the values of x that satisfy $x + 3 = 5$, one calls such expressions open sentences and asks for the truth set of this open sentence. The truth set means the values of x which make the equation correct. Of course, there is only the value $x = 2$ in the truth set of this open sentence. The truth set of the open sentence $x^2 = 4$ consists of 2 and -2.

To secure precision the modernists have replaced many definitions in the traditional texts with their own versions. In a traditional text a variable might be defined as a symbol or letter which can take on any one of some collection or set of values. Thus the x in $y = x^2$ can be assigned any real number. This language is not acceptable in modern mathematics. Instead, a modern text says that a variable is a symbol which may represent any of the elements of a specified set. A set whose elements serve as replacements for a variable is called the replacement set for the variable. This set is also called the domain of the variable. The individual members of the replacement set are called the values of the variable. A variable with just one value is called a constant.*

Variables are important because they enter into functions. Thus $y = x^2$ is a function, and the traditional definition of a function is a relationship between variables such that if a value is assigned to one variable the value or values of the second one are determined. Thus, in the case of $y = x^2$, if $x = 3$ then y has the single value of 9. On the other hand, if the function is $y^2 = x + 5$, then when $x = 4$, $y = +3$ and $y = -3$. The function $y = x^2$ is said to be single-valued and the function $y^2 = x + 5$ is said to be multiple-valued.

Such "sloppiness" is not permitted in the modern texts. They first introduce the notion of an ordered pair. Thus $(3,4)$, $(5,6)$, and $(6, -2)$ are ordered pairs of real numbers. The concept of function, single- or multiple-valued, is replaced by the concept of a relation. A relation is any set of ordered pairs. A function (meaning single-valued function) is a relation in which no two different ordered pairs have the same first number. Thus, $(4,3)$ and $(4, -3)$ could not belong to the set of ordered pairs which define a function. Given these definitions of relation and function, the students are expected to see that $y^2 = x + 5$ is a relation and $y = x^2$ is a function.

Precision is secured in geometry by carefully distinguishing concepts. An angle, for Euclid, is the inclination to one another of two lines which meet. Apparently this will not do. An angle now is the figure formed by two rays (half-lines) that meet in a common point O (Fig.

---

* Logically this definition of a variable seems troublesome. If a constant must be defined as a variable with just one value, then what are the elements of the replacement set? Are they not constants? If so, the concept of constant is already involved in the definition of a variable.

6.1). Of course, if an angle is just the two rays we do know which angle is referred to, angle A or angle B. To decide we need to know what the interior of an angle is. For this purpose we call upon another axiom (or theorem in some developments). This axiom states that a

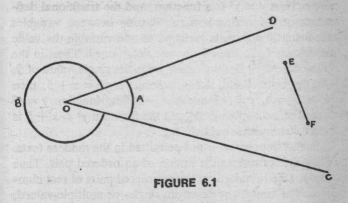

FIGURE 6.1

line divides the plane into two parts. Hence, if we take the part determined by the line OC (which contains the ray OC) and containing D and take the part determined by the line OD and containing C and now take the set intersection of these two parts (the points common to these two parts) we obtain the interior of the angle DOC. Having determined the interior of the angle we can prove that if E and F are in the interior of an angle (Fig. 6.1), then the segment EF lies in the interior of the angle.

Knowing which angle we have in mind does not tell us what the measure (size) of the angle is. For this purpose another axiom tells us in effect that to all rays emanating from O (Fig. 6.1) and beginning with OC numbers can be assigned and the number assigned to OD is the measure of angle DOC. Thus to OC the number 0 is assigned because the numbering starts with it and to OD the number 30 might be assigned. Then the size of angle COD is 30.

Angles are often studied as parts of triangles. We might be tempted to say in looking at Figure 6.2 that angle *A* is an angle of triangle *BAC* and that *BA* and *CA* are the sides of angle *A*. This would be grossly incorrect. The sides of an angle are rays which extend to infinity, whereas *BA* and *CA* are finite segments. However, some authors grant dispensation to favored readers and permit them to speak of angle *A* as an angle of triangle *BAC*.

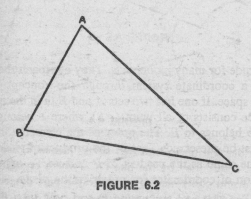

FIGURE 6.2

Euclid was careless in defining a triangle as a figure formed by three line segments. Of course this will not do. Figure 6.3 is formed by three segments and is not a triangle. "Properly," a triangle is the figure consisting of the union (a set theoretical notion) of three noncollinear points and the line segments joining these points.

In elementary algebra and again in coordinate or analytic geometry students learn what a rectangular coordinate system is. In this system each point of the plane is located by two numbers. The first is the distance of the point to the right or left of the *y*-axis and the second is the distance above or below the *x*-axis. Thus the coordinates of *P* in Figure 6.4 are 2 and 4. These are written as the ordered pair (2,4). This approach to coordinates

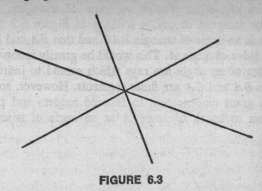

FIGURE 6.3

is too crude for many modernists. They approach the notion of a coordinate system through the concept of a product space. If one has two sets A and B then the product space consists of all pairs (a,b) where a belongs to A and b belongs to B. The order of the elements in the pair must be preserved. That is, the product of A and B is not the same as the product of B and A. To arrive at the notion of coordinates for points in the plane, A and

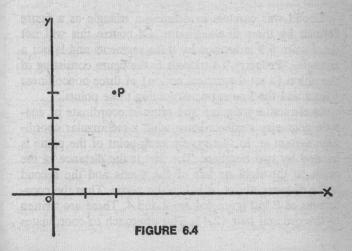

FIGURE 6.4

B are each taken to be the set of all real numbers and the product space of these two sets is the set of coordinates (*a,b*). Then *a* is identified with the distance of *P* from the *y*-axis and *b* with the distance of *P* from the *x*-axis. The notion of product space is supposed to introduce precision in the definition of coordinates.

In keeping with their aim to secure precision, the modern texts define carefully every concept that is used. The consequence is an immense amount of terminology. Thus one finds definitions for angle, triangle, polygon, numeral, equation, open phrase, open sentence, compound sentence, algebraic expression, binary operation, closure, inverse, uniqueness of inverse, null set, union, intersection, solution set, line segment, point pair, distance, length, ray and many other terms. An actual count of the number of terms introduced in the ninth-grade algebra course and tenth-grade geometry course shows that several hundred terms are introduced in each one. Of course, students are expected to learn and use these terms.

Much of the terminology is abstract. The authors of the modern texts naturally wish to introduce addition and multiplication. These operations applied to numbers are taught in the elementary grades. Applied to letters they are taught in algebra. However, the terms addition and multiplication are considered too mundane. To treat adding 5 and 7, the texts speak of a binary operation, meaning that addition applies to two numbers or letters. Multiplication, likewise, is a binary operation.

These examples may illustrate what the modernists mean when they say that errors and fuzzy thinking can be eliminated by precise language. Now we'll consider their claim. Typical of the precision is the distinction we have already noted, namely, the distinction between a number and a numeral. The distinction raises a question. If 7 is a numeral for the number seven, what is the number? This, the students are told, is an idea in their minds. The answer is hardly satisfying to young people and the distinction has done more harm than good. It has shrouded number in mystery and made students more dubious of their capacity to understand.

Of course, there is a distinction between number and numeral, and between Robert Smith, the name, and Robert Smith, the boy. However, language incorporates practices that may cut corners but nevertheless promote effective communication. Excess verbiage is to be avoided if there is no danger of error. It would seem unnecessary to say "the boy whose name is Robert Smith" when saying Robert Smith clearly refers to the boy and not to his name. Context determines meaning in almost all cases. The 2 in 245 and in 425 have different meanings; yet no confusion is incurred.

Much of the new terminology is totally unnecessary. To speak of binary operations, closure, and many, many other terms labels what need not be labeled. Some terminology replaces older terminology but to no particular advantage. Thus $x + 2$ had been called an expression. It is now an open phrase, open because the value of x is not specified. An equation, say, $x + 2 = 0$, will presumably become clearer if it is renamed an open sentence. The problem of solving an equation in x becomes the problem of finding the truth set of an open sentence, but the job of solving an equation remains precisely what it was before the new terminology was introduced.

The common understandings which students have acquired through experience are good enough; formal definitions are not usually needed. Students know what a triangle is and do not have to be taught that it consists of the union of three noncollinear points and the line segments joining them. After reading such a definition one must think long and hard to realize that it defines the familiar triangle.

The terms "relation" and "function" have replaced multiple-valued and single-valued functions. The older term, multiple-valued function, was meaningful; the newer one, relation, is vague. Beyond this, relation and function are defined in terms of sets of ordered pairs. Such a set is supposed to convey the idea of, say, $y = 3x$, wherein x can take on all real numbers as values and y is three times x. However, the function has an infinite number of x and y values. The definition as a set of ordered pairs gives a distorted view because it gives the impression that the number of pairs in the set is finite. Further, the function cannot be specified by enumerating the set of ordered pairs because the number of these is infinite whereas $y = 3x$ does convey the entire function. Finally, the basic intuitive idea of a function—that is, as the values of one variable, x, change, the values of the second variable, y, are dependent on the values of the first one and change with it—is lost in the static description of a set of ordered pairs. Apparently novelty is desirable even at the cost of understanding.

It is only fair to admit that there are functions wherein x may take on only a finite number of integral values. For example, the function describing the population of, say, New York City, wherein x is the time, say the number of months from January 1, 1950, to January 1, 1970, and y the corresponding population, will have only a finite number of integral values for x and for y. Though the ordered pair definition is more applicable in these special cases, it is not helpful.

The introduction of so many new terms, and particularly terms which are not suggestive of the concepts they represent, puts an intolerable burden on the memory. The excessive terminology has been criticized by Richard P. Feynman, professor of physics at the California Institute of Technology, and a Nobel prize winner in 1965. Professor Feynman served as a member of the California

State Curriculum Commission which examined texts to be used in the California schools. In his article "New Textbooks for the New Mathematics," which appeared in *Engineering and Science,* he attacked the excessive terminology introduced in geometry. "Some of the books go a long way with the definition of a closed curve, open curve, closed regions, and open regions and so on, . . . and yet they teach no more geometry than the fact that a straight line drawn in the plane divides the plane into two pieces. At the end of some of these geometry books, look over to find, at the end of a long discourse, or a long effort at learning, just what knowledge of geometry has been acquired. I think that often the total number of facts that are learned is quite small, while the total number of words is very great. This is unsatisfactory. Furthermore, there is a tendency in some of the books to use most peculiar words—the words that are used in the most technical jargon of the pure mathematician. I see no reason for this."

Terminology, particularly pretentious terminology, is no substitute for substance. In view of the emphasis on terminology the reformers evidently believe that giving names to things automatically confers power over them. Many critics have charged that the modern mathematics texts are no more than dictionaries or studies in linguistics.

There is little doubt that the newness attributed to the new mathematics results largely from the introduction of a new terminology which serves less well than the older one. What has been brought into modern mathematics is not so much modern mathematics as the verbiage and sometimes the parody of it.

Another aid to precision heavily exploited by the new mathematics is symbolism. Since sets are a basic idea in this curriculum, there is a notation for them. Thus {1,2,3} denotes the set containing the objects 1, 2 and 3.

This curriculum also distinguishes between an object and the set which consists of that object. Thus 3 and {3} are different entities. The reason for the distinction is that the set {3} can be added to other sets, but the object 3 cannot be added to other sets, just as horses and cows cannot be added unless we think of them as belonging to a higher genus of animals. Whether the students appreciate this distinction is open to question; but if they do, the next point is likely to upset them. In the theory of sets the concept of an empty set enters. The empty set, commonly denoted by the Greek Ø, is illustrated by the set of all kings of the United States. However, the set consisting of the empty set, namely {Ø}, is not empty because it contains the empty set.*

To specify the set of all values of x which satisfy $x + 2 = 4$, that is, the values which make true sentences of the open sentence $x + 2 = 4$, the modern notation calls for $\{x | x + 2 = 4\}$. To indicate that 1 is in the set, 1,2,3 one writes $1 \in \{1,2,3\}$, the symbol $\in$ meaning "belongs to." To designate properly the collection of all dogs in the United States one must write $\{x | x$ is a dog in the United States$\}$.

With the above symbolism one can frame more complicated statements in symbolic form. Thus to specify the collection of all cats and dogs in the United States one writes $\{x | x$ is a dog in the United States$\} \cup \{x | x$ is a cat in the United States$\}$. Here $\cup$ stands for the union of two sets, that is, the collection of all objects that are in at least one of the two sets. If one wishes to ask the question, which of the numbers 1,2, and 3 satisfy the inequality $3x - 1 < 8$, one must now write, if $x \in \{1,2,3\}$ determine the elements of $\{x | 3x - 1 < 8\}$. Similarly to describe the values of x and y for which $x + y = 5$, one writes $\{(x,y) | x + y = 5\}$. The positive values of y in

---

* Thus we can build something out of nothing and rise from rags to riches.

$x - 6y = 10$ that correspond to $x = -10$ and $x = 2$ are denoted by the solution set for $x - 6y = 10$, $x \in \{-10, 2\}$, $y \in \{\text{positive number}\}$. To specify that there is at least one value of x for which $x + 2 = 3$ one uses the notation $\{\exists x | x + 2 = 3\}$. On the other hand, to state that no even number is odd, one lets p_x stand for x is even and q_x for x is not odd. Then using the symbol $\rightarrow$, which means implies, one writes $V_x(p_x \rightarrow q_x)$, which states in words that for all x, x is even implies that x is not odd. The symbols $\exists$ and V are called quantifiers.

The criticisms that apply to terminology apply equally to the use of symbols. It is of course granted that some symbolism is useful and even necessary. Well-motivated and suitably chosen notation contributes to the clarification of essential mathematical concepts and relations and saves labor in operations. It also aids in the understanding of ideas. To say in words what the expression $a^2 + 2ab + b^2$ states would not only require more length but would make comprehension difficult.

However, in the excessive use of symbols the modern mathematics curriculum has made a vice out of a virtue. Consider the following example. Given the expression $f(x,y) = 3x - 2y = 1$, determine the set

$$A = \{(x,y) | x \in N, y \in N, f(x,y) = 1\},$$

where N means the set of all integers. All this symbolism amounts to the statement, Determine all the integral solutions of $3x - 2y = 1$. The modern authors are symbol-happy. Thus we find braces, brackets, vertical bars, parentheses, quantifiers, cup and cap, the one-way implication symbol and the two-way implication symbol, ϵ for belongs to, and many other symbols. Students are stunned by dark forbidding symbols.

Many symbols serve almost no purpose; the English language is better. The slight saving in space is more than offset by the psychological handicap that symbolism imposes on the students. To wallow in symbols is to make reading and comprehension more difficult. When the burden of remembering what the symbols stand for becomes great, more harm is done than by using verbal statements. Moreover, symbols frighten students and so should be used sparingly. The difficulty in remembering the meanings and the general unattractiveness of symbolic expressions repel and disturb students; the symbols are like hostile standards floating over a seemingly impregnable citadel. The very fact that symbolism entered mathematics to any significant extent as late as the sixteenth and seventeenth centuries indicates that it does not come readily to people.

Symbolism can serve three purposes. It can communicate ideas effectively; it can conceal ideas; and it can conceal the absence of ideas. It often seems as though the modern mathematics texts use symbolism to conceal the poverty of ideas. Alternatively the purpose of their symbolism seems to be to make the obvious inscrutable and so repel the understanding.

The inordinate emphasis on symbolism would give most people an impression of mathematics analogous to that derived from a presentation of music which puts the total effort to learning to write and read musical notation. No inkling would be given of what the notes, sharps, flats, the beats in a measure or any other symbols mean in terms of actual sounds, beautiful themes and complete compositions which the symbols merely record. Indeed, the analogy between mathematics and music, so far as education is concerned, extends even further. No one writes out the notes of a musical composition and then plays it to see what the notes call for. The ideas and even the entire development are envisioned and "played" in the composer's mind before he records them in the musical notation. So too the ideas and the arguments with which the mathematician is concerned have physical, intuitive or geometrical reality long before they are recorded in the symbolism. One sees then that the symbols of mathematics, like the notes of music, are in themselves merely an artificial, intrinsically meaningless script. They will convey life, meaning, richness of thought and beauty only if the ideas and the thinking which the symbols merely record are taught with as little use of symbolism as possible.

Despite the disadvantages in the use of symbols, the modern mathematics texts prefer to use them generously. One suspects that they do so to give an air of profundity to simple and straightforward material. One even finds verbal statements "elucidated by symbolic expressions," as though symbols clarify words.

The ridiculousness of the efforts to secure precision through terminology and symbolism has been attacked by Professor Feynman. In his article "New Textbooks for the New Mathematics," he criticized the precision sought by using set language. He mimics the precision by pointing out, "A zookeeper, instructing his assistant to take the sick lizards out of the cage, could say, 'Take that set of animals which is the intersection of the set of lizards with the set of sick animals out of the cage.' This language is correct, precise, set theoretical language, but it says no more than 'Take the sick lizards out of the cage.' . . . People who use mathematics in science, engineering, and so on, never use the long sentences of our imaginary zookeeper. . . . It will perhaps surprise most people who have studied these textbooks to discover that the symbol ∪ or ∩ representing union and intersection of sets and the special use of brackets and so forth, all the elaborate notation for sets that is given in these books, almost never appear in any writings in theoretical physics, in engineering, in business arithmetic, computer design, or other places where mathematics is being used. I see no need or reason for all this to be explained or to be taught in school. It is not a useful way to express one's self. It is not a cogent and simple way. It is claimed to be precise, but precise for what purpose?"

Feynman includes in his critique these words: "Many of the math books that are suggested now are full of such nonsense—of carefully and precisely defined special words that are used by pure mathematicians in their most subtle and difficult analyses, and are used by nobody else. . . . The real problem in speech is not *precise* language. The problem is *clear* language." He gives as an example a now common form of attempted precision, the distinction between the picture of a ball and a ball. A text says, "Color the picture of the ball red" instead of "Color the ball red." Feynman points out that the phraseology "Color the picture of the ball red" begins to produce doubts, whereas "Color the ball red" would not. The picture of the ball includes the ball and a background. Should one color the entire background too? As Feynman also indicates, the material on sets, now given heavy play, is used only by forcing it into artificial and complicated constructions.

7

Mathematics for Mathematics' Sake

"One may say that mathematics talks about
things which are of no concern at all to man.
. . . It seems an irony of creation that man's
mind knows how to handle things the better the
farther removed they are from the center of his
existence. Thus we are cleverest where
knowledge matters least: in mathematics,
especially in number theory."

Hermann Weyl

The new mathematics presents the subject as self-sufficient. Presumably mathematics can feed on itself to
grow and offers values when studied in and for itself.

For example, mathematics is presented as self-generating. Thus, granted the whole numbers, the fractions
can be and are introduced in the new curricula by asking
for some number x that solves $3x = 7$. Clearly no whole
number will do and therefore, so the story goes, mathematicians are led to introduce fractions such as 7/3.

Given the positive whole numbers and fractions one asks what number x satisfies the equation $x + 5 = 2$. Again the existing numbers are found to be useless for solving such equations and so negative numbers are created. The same approach motivates the introduction of irrational numbers and complex numbers, that is, numbers involving the square root of negative numbers, for example, $\sqrt{-5}$.

Not only are the various types of numbers introduced by raising mathematical questions, but the axioms that hold for the whole numbers are assumed to apply to each new class introduced and thereby one proves properties of each of these classes. We have illustrated this latter fact in connection with the deductive approach adopted by the new mathematics.

The introduction of new mathematical ideas by raising questions about old ones is illustrated by another situation. Having solved linear equations, that is, equations of the form $ax + b = 0$, it is mathematically relevant to ask whether one can solve equations such as $x^2 + 7x + 9 = 0$, $x^2 - 5x + 4 = 0$, and more generally $ax^2 + bx + c = 0$.

Not all of mathematics can be arrived at by raising questions about ideas already studied. Geometry must be introduced afresh. However, once this subject is launched, it is easy to raise mathematical questions that lead to new geometrical topics. For example, after students have studied congruent triangles one can ask, "What can be said about two triangles for which the angles of one equal the angles of the other?" Such triangles need not be congruent but they are similar, and thus the study of similar triangles is suggested. Again, having considered triangles, which are three-sided figures, one can raise and answer questions about four-sided figures or quadrilaterals. These examples may serve to illustrate what is meant by mathematics being self-generating. New concepts and new problems are introduced by raising questions about concepts already studied.

Historically this approach to mathematics is certainly false. The significant concepts, operations, theorems, and even methods of proof were suggested by real situations and phenomena. Mathematics grew out of our experiences in the physical world. For example, as we have already noted in another connection, our method of adding fractions was adopted because the sum so obtained represents what results physically when fractional entities are put together. We could instead add the numerators and add the denominators so that $1/2 + 1/2 = 2/4$, but this result does not apply when $1/2$ of a pie is added to $1/2$ of a pie. Likewise, the properties which mathematical operations possess are not dictated by extending the associative, commutative and distributive laws to the new elements. If matrix multiplication, to be useful, must be noncommutative we abandon commutativity, though we could define a commutative multiplication of matrices. Hence, extending the associative, commutative and distributive axioms to new classes of numbers can lead to useless mathematics. Geometry too arose out of the study of real figures existing in physical space and the desire to know the properties of these real figures and space itself.

The historical origins of mathematical concepts and processes need not of course be the pedagogical approach. However, a valid objection to generating new concepts and operations through older ones is the meaninglessness of what is introduced. For example, to introduce negative numbers some modern texts ask, "What number added to 2 gives 0?" They then introduce -2 as the requisite number. As some texts put it, -2 is the unique additive inverse to 2. But this introduction of -2 gives no more understanding than the statement, "Antimatter is that substance which added to matter produces a vacuum," gives any understanding of antimatter.

By generating mathematics through mathematical questions and by extending to new domains laws or axioms that hold in previously established ones, mathematics is isolated from all other bodies of knowledge. It exists for its own sake and is presumably self-sufficient. It then appears that by chance the deductive structures so erected fit some physical phenomena and mathematics can be applied to real problems. However, this seemingly fortuitous value is not utilized. Mathematics in the modern mathematics texts is not applied to real problems. Some authors make minor concessions to applications in the seventh- and eighth-grade material. Having eased their consciences they ignore applications in the higher grades.

The isolation from the real world is evident from the artificial problems found in the texts. Beyond the purely technical exercises which serve as drill and which certainly have no connection with the real world, one finds exercises whose character is illustrated by the following.

Dividing a certain number by two yields the same result as subtracting fifteen from three times the number. Find the number.

Harry earns three times as much per week as Tom does, while Dick earns eighty dollars a week more than Tom. If Dick and Harry have the same salary, how much does each of the three men earn?

Bill is twice as old as Mary. If he is exactly ten years older than Mary was last year, how old are Bill and Mary?

The Red Cross knitted fifty sweaters in ten days. The Junior Red Cross assisted, contributing two dozen fewer than the senior organization. How many did the Junior Red Cross knit?

How large is an angle whose supplement contains 21° less than four times its complement?

Here is a "science" problem: According to the Law of Reflection: $i = r$. Given that $i = (2n + 30)°$ and that $r = (4n - 10)°$, find n. Not a word is included about the meaning of the Law of Reflection. It could refer to mental reflection.

The artificiality of the problems is obvious. Their inanity and pointlessness should make any sensible student writhe with mental pain. Of course, these artificial problems have been used in the traditional curriculum also.

One would have thought from the pronouncements of several curriculum groups that real applications would have been incorporated and even stressed at least in some of the new programs. For example, the Commission on Mathematics in its 1959 report, *Program for College Preparatory Mathematics,* justified the need for a new curriculum by pointing out that there were many new uses of mathematics in fields such as the exploration of space, nuclear science, the social sciences, psychology, business, and industry. Other groups echoed this thought. But no such applications have been included, to say nothing about the older applications to the physical sciences.

The neglect of applications has been noted and deplored even by some advocates of the new mathematics. Thus Professor Paul Rosenbloom, one of the active workers on the new curriculum, said in an article on applied mathematics (see the reference to Carrier in the bibliography), "In the writing teams for grades 9 to 11, some of us were disappointed that the people we thought have been advocating applications came up with writing stuff about open sentences and the like." But the applications have not been supplied.

One reason is that the professors and teachers who formulated the new curricula do not know science. The professors were pure mathematicians and the courses taken by teachers were offered largely by these professors, so that the teachers too are ignorant of what mathematics can accomplish. Both groups therefore even prefer a treatment which is purely mathematical because it avoids presenting and explaining a few physical concepts and how to formulate physical problems mathematically.

The modernists apparently also want to keep their subject pure. They don't wish to sully it; they desire to remove the dross of earth from which mathematics has risen. But as they wash the ore they keep the iron and lose the gold. A perfect command of the English language is useless if a man has nothing to say, and pure mathematics has little to say to young students. As Bertrand Russell put it, "Mathematics may be defined as the subject in which we never know what we are talking about nor whether what we are saying is true." Though Russell had in mind the logical structure of mathematics his statement describes what is being taught. The contents and spirit of the modern mathematics curricula may suit the mathematical scholar but the relation to the real world has been ignored.

Of course mathematics is not an isolated self-sufficient body of knowledge. It exists primarily to help man understand and master the physical world and, to some extent, the economic and social worlds. Mathematics serves ends and purposes. If the subject did not have these values it would not get any place at all in the school program. It is because mathematics is extraordinarily helpful that it is in great demand and receives so much emphasis today. These values should be reflected in the curriculum.

During the last few years many of the curriculum leaders acknowledged that they had neglected to point out the applications of mathematics. But their approach to remedying this deficiency is ludicrous. They call upon applied mathematicians of some large research laboratories or industrial organizations to supply applications. These men abstract from genuine applications snippets of mathematics that are indeed involved in applications. But the snippets reveal nothing of what is accomplished. They are like salt in a cake. The students are asked to eat the salt in the expectation that they will thereby enjoy the cake.

The isolation of mathematics was attacked by Dr. Alvin M. Weinberg in his article "But Is the Teacher Also a Citizen?" Dr. Weinberg, director of the Oak Ridge National Laboratory, charged that professors are so much absorbed in their own disciplines that they are neglecting to teach students useful knowledge, knowledge that can be applied to other fields, as mathematics is applied to physics, and knowledge that would make them more useful members of society. The new mathematics he terms "a puristic monster."

To present mathematics as self-generating is not only a denial of history but conceals its vital connections with other branches of knowledge. From a pedagogical standpoint this approach is most unfortunate because it foregoes the opportunity and great need to give motivation and meaning to mathematics. Since the ideas of elementary mathematics did arise from physical and practical problems, these very problems or more modern equivalent ones could be used to motivate the study of mathematics. We have already pointed out that the gravest defect of the traditional curriculum is the lack of motivation. Instead of remedying this most serious defect the modern curriculum has aggravated it. One cannot motivate young people to learn mathematics with more mathematics. Students who are not interested in solving $x + 3 = 4$ will certainly not be interested in solving $x + 4 = 3$.

To isolate mathematics is also to rob it of meaning.

The student may be persuaded to study the mathematical function $s = 16t^2$. But the function as such has no meaning. Physically it does represent the motion of a ball that is dropped. The variable s represents the distance fallen in t seconds. Given this interpretation the student can visualize s and t increasing as the ball falls, and with a little help he can appreciate how this function differs from $s = 16t$ or $s = 16t^3$. Indeed, the physical appreciation of how the various functions differ is the surest way to convey understanding of the nature and behavior of these functions.

By isolating the subject, mathematics becomes pointless and unattractive. It is as if the subject were taught in a room with mirrors on the walls rather than with windows to the outside world. The modern mathematics leaders have assumed that mathematics taken in and for itself is attractive to young people. But this is hardly true. To the student mathematics proper appears as no more than a huge intricate puzzle. Pure mathematics may provide challenging problems—but so do law, economics, and biology, and these subjects appear far more vital and relevant to the students. Mathematics, by reasons of its abstractness, is not a natural human interest. The very fact that only a few civilizations among hundreds have devoted some time and effort to the subject may show how unnatural the subject is.

The modern mathematics advocates are of the opinion that students will find values in studying mathematics for its own sake. One of the values they extol is structure. Indeed structure has become the fashionable word. Thus the Preface to the course in matrix algebra fashioned by the School Mathematics Study Group states, "The discernment of structure is essential, no less to the appreciation of painting or a symphony than in the behavior of a physical system; no less in economics than in astronomy." Again in its Preface to the Teachers

Commentary on its first-year algebra course, this group states, "The principal objective of this First Course in Algebra is to help the student develop an understanding and appreciation of some of the algebraic structure exhibited by the real number system and the use of this structure as a basis for the techniques of algebra."

Just what is structure? Basically any branch of mathematics consists of definitions, axioms and theorems. This is structure in the large. However, the properties that hold in one branch may not hold in another. Thus multiplication of any two real numbers is commutative but the multiplication of matrices is not. Consequently the logical structure of the real numbers is different from that of matrices.

What can young students be taught about structure? Given the positive whole numbers, and zero—that is, the counting numbers—it is true that the associative and commutative properties hold for addition and multiplication. However, one cannot subtract 5 from 2 and remain in the class of the counting numbers. Alternatively one can say that 3 has no inverse in the class of the positive whole numbers; that is, there is no positive number a such that $3 + a = 0$. On the other hand, in the class of the positive and negative whole numbers, 3 does have an inverse, namely, -3. Hence the positive and negative whole numbers possess a property that the positive whole numbers alone do not possess. Thus the two classes of numbers differ in structure.

It is not possible in the class of the positive and negative whole numbers to divide every number by another. Thus $1/7$ does not exist in this class. Alternatively one can say that there is no inverse to multiplication in the class of the whole numbers; that is, there is no number x such that $7 \times x = 1$. On the other hand, in the class of positive and negative whole numbers and fractions, each number does have an inverse with respect to multiplication. Hence in this class every number has an inverse with respect to both addition and multiplication, so that the structure of the rational numbers (positive and negative whole numbers and fractions) differs from that of the whole numbers.

Since the high school students do not go much, if at all, beyond the use of real numbers, that is, rational and irrational numbers, they do not have occasion to learn many structures or the opportunity to learn more diverse structures. Nor can they compare many structures as to similarities and differences. Nevertheless, any number of proponents of modern mathematics stress the learning of structure. Beyond the quotations already given we find the Commission on Mathematics stating in its *Program for College Preparatory Mathematics* (p. 2): "The contemporary point of view, while not discounting the manipulative skills necessary for efficient mathematical thought, puts chief emphasis on the structure or pattern of the system and on deductive thinking." Again in the pamphlet *The Revolution in School Mathematics,* published by the National Council of Teachers of Mathematics, Kenneth E. Brown of the U.S. Department of Health, Education, and Welfare, states, "Another area of emphasis common to all improved programs is structure. It is reflected in the careful development of mathematics as a deductive system."

There is nothing intrinsically wrong with the goal of teaching structure, though one might question its importance at the elementary stage of mathematical learning. However, the possibility of teaching structure meaningfully is certainly in question. To appreciate the differences and similarities in the structure of the members of our physical world one must meet and know well a great variety of animals. One who knows only cats and dogs will readily believe that all animals have the same structure and in fact it will not even occur to him to think about structure. If, however, he meets giraffes, elephants, fish, and birds the subject of structure may strike him as worthy of investigation.

The elementary and high school student is in the position of a man who knows only cats and dogs. It is true that one can differentiate between the logical structure of the whole numbers and the signed integers and between the signed integers and the rational numbers. However, the student is still struggling to understand these numbers and the operations with them and is not prepared to take the overall view which the appreciation of structure requires. Even if he glimpses some differences in the operations permissible with rational numbers as opposed to the integers, he is not likely to be impressed with the concept of structure. If he continues with mathematics and encounters algebras wherein multiplication is not commutative, he may then begin to take notice of differences in structure. It is unrealistic to expect people who have seen only doghouses and pigpens to appreciate architecture.

Another of the claims made for the teaching of structure is that it unifies a body of mathematics because it shows that the theorems all follow from one set of axioms and are arranged in a logical sequence. But this facet of structure has very little value to high school students. Elementary algebra in the modern mathematics curriculum remains the usual hodgepodge of disconnected topics that it is in the traditional curriculum. The fact that all of these topics can be treated from one basic set of axioms may give them unity in the mind of a mathematician but this is hardly a unifying or impressive connection especially to young people who have yet to learn what a deductive structure is.

Mathematics teachers do often talk about giving the students a feeling for the power of mathematics, and they also speak of doing it through exhibiting the structure and order which permeate every branch of the subject. Just how these features make evident the power of mathematics is not clear. To demonstrate this power one must use it in real situations. This is where the power is applied and this is how students will get to appreciate it.

Thus the core of the criticism of teaching structure to young people, aside from its importance or lack of importance, is that the subject cannot be significant at this stage. And this very fact implies that it should not be taught at this level.

8

The New Contents of the New Mathematics

"Wisdom oft is nearer when we stoop than
when we soar."

William Wordsworth

Logical development as the road to understanding, rigor,
precision through terminology and symbolism, and emphasis on mathematics for its own sake are all employed
in the modern mathematics approach to the curriculum.
What subject matter is favored? The old subjects, arithmetic, algebra, Euclidean geometry, trigonometry, and
the elements of analytic geometry are still taught in the
new curriculum, despite the claim made by many modernists that this pre-1700 mathematics is outmoded and
even useless in modern society. Of course, the amount
of traditional material taught varies somewhat from one
version to another of the modern mathematics curriculum
but it is the predominant part in every case.

However, the new curriculum does offer some new
contents. By far the most emphasized among the new
topics is set theory. This subject is now taught from the
kindergarten up, as though students would starve, men-

tally at least, if they did not have this diet. A set, as any modern mathematics text will tell us, is no more than a class or collection of objects. A set of apples, a set of garbage cans, a set of letters of the English alphabet, and the set of natural numbers are examples. The concept and the word *set* are simple enough. However, *set theory* contains many subtle concepts and theorems. The two basic concepts are the union of two sets and the intersection of two sets. By the union of the set of red objects and the set of books we mean the set of all objects that are either red or are books. The intersection of these two sets is the set of all objects common to the two. Thus, since a book with red covers is a red object, then the set of red books is the intersection of the two sets. One can now speak of the union and intersection of three or more sets and combinations of unions and intersections. A set usually has subsets. Red chairs would be a subset of all red objects. One subset is contained in every set and this is the empty set. This set could stand for the set of women presidents of the United States.

The most significant sets are infinite sets. Thus the set of natural numbers is infinite. Students are taught that two sets are equivalent if it is possible to set up a one-to-one correspondence between them. By means of the correspondence

there is a one-to-one correspondence between the set of natural numbers and the set of even natural numbers, in that each natural number corresponds to double itself and conversely. These two sets and indeed any set that can be put into one-to-one correspondence with the natural numbers are said to have the same number of objects. So there are as many even numbers as natural numbers despite the fact that the even numbers are only a part of the set of natural numbers. Students are then shown that the set of rational numbers and the set of natural numbers can be put into one-to-one correspondence so that there are as many rational numbers as natural numbers. They are also taught that the set of real numbers cannot be put into one-to-one correspondence with the natural numbers, and since the set of real numbers contains the natural numbers it is a larger set than the set of natural numbers.

The emphasis on sets is justified by the modern mathematics proponents on several grounds. The first is that it is a basic concept of mathematics. Thus, numbers are names for sets of objects (though the union and intersection of sets are not the same operations as addition and multiplication of whole numbers). The second contention is that the concept of set unifies various branches of mathematics. Thus the notion of set is used to speak of a solution set for the roots of equations, to define geometrical figures, and to define relation and function in terms of sets of ordered pairs of numbers. The unification through sets, at least on the elementary level, is limited to special terminology and a questionable refashioning of previously accepted and acceptable definitions of concepts.

Perhaps the second most popular topic of modern mathematics is bases of number systems. This concept dates back to the Babylonians of 2500 B.C. who used base sixty. The concept was thoroughly aired by the famous mathematicians John Wallis and Gottfried Wilhelm Leibniz in the seventeenth century.

Just what is a base? Our method of writing quantities presupposes base ten. Thus 372 means $3 \times 10^2 + 7 \times 10 + 2$. The same quantity can be written in another base, say eight. Since $372 = 5 \times 8^2 + 6 \times 8 + 4$, 372 written in base eight is 564. Students are taught to write numbers in other bases and how to add and multiply in these bases. It so happens that modern electronic computers operate in base two. One might expect, then, that students would be taught bases when they are about to learn about computers. But the contention of the modern mathematics proponents is that learning how to operate in bases other than ten aids the understanding of base ten and the operations of arithmetic.

A third common topic in the new mathematics is the study of congruences. This subject is frequently introduced by what is called clock arithmetic. Our clocks record up to twelve and then start over again with zero. Thus if twenty-two hours have passed from twelve o'clock, the clock will not read 22 but 10. This suggests that all numbers be reduced by as many twelves as can be subtracted from them. Twenty-two will be reduced to ten and is said to be congruent to ten modulo twelve. To make the arithmetic simpler students are taught congruences modulo five or modulo six. Now, congruences have no application to science or engineering. They are taught for their mathematical interest and, as a matter of fact, the topic belongs to the theory of numbers, which is a subject pursued primarily for its own sake. Nevertheless it is a curiosity and may awaken some interest in numbers. One does have to make up new addition and multiplication tables. Thus modulo 6, $4 \times 3 = 0$ because if one takes the normal product of twelve and subtracts as many sixes as possible, one obtains zero. This product also illustrates another curious feature, namely, that the product of two numbers can be zero, though neither factor is zero.

Another topic favored by the modern mathematics curriculum is inequalities. A simple example would call for the values of x for which $3x < 6$. This topic has been taught in the traditional college algebra for many generations, but the new curriculum has moved it down to ninth-grade algebra.

The subject recommended for the second semester of the twelfth grade by the School Mathematics Study Group is matrices. A matrix is a rectangular, usually square, array of numbers. Thus

$$\begin{pmatrix} 2 & 3 \\ -5 & 7 \end{pmatrix}$$

is a matrix and is said to be of second order because it has two rows and two columns. Such matrices can be added, subtracted, multiplied and usually divided by one another. They can also be converted into matrices with different numbers by multiplying them by other matrices. In other words, there is an algebra of matrices.

Many modern curricula teach symbolic logic. In ordinary reasoning we combine statements in various ways. Thus one might say, "It is not raining" and "I am going for a walk." Here the two statements, "It is not raining" and "I am going for a walk" are independent propositions connected by the conjunction "and." The joint assertion of these two propositions p and q is evidently true if and only if both propositions are true. However, this evident meaning is not accepted as such but is defined by what is called a truth table which looks as follows:

p	q	p ∧ q
T	T	T
T	F	F
F	T	F
F	F	F

The symbol $\wedge$ means "and"; the truth table tells us that the joint assertion is true if and only if p and q are both true. Presumably the truth table is more informative and more precise as a definition of "and" than the verbal statement. There are similar truth tables for "or," "implication," and "negation." These truth tables are then used to prove theorems of logic. Thus one proves by this means that the negation of the assertion "It is not raining and I am going for a walk" means either that it is raining or I am not going for a walk or both. One is supposed to learn how to reason through the use of these truth tables.

Many modern texts teach Boolean algebra which is also intended to aid in reasoning and is an alternative to the use of truth tables. The algebra is the same as the algebra of sets. For example, the set of dogs plus the set of dogs is just the set of dogs in set theory. In Boolean algebra if a represents the set of dogs, the algebraic statement is $a + a = a$. If b represents the set of animals, the statement that all dogs are animals is represented by $ab = a$, wherein ab means the set of objects in a and in b. By using Boolean algebra one can perform ordinary reasoning in purely symbolic form.

Modern mathematics texts favor abstract concepts. Before students have worked with functions they are asked to learn about relations and functions in terms of sets of ordered pairs (Chapter 6). It is only after the general definitions are taught that students learn to work with $y = x^2$, $y = x^3$, and the like.

To get the students to practice abstraction they are asked to answer exercises such as the following: If ϕ is an operation on the positive numbers, for which one of the following definitions of ϕ is $x \phi y = y \phi x$:

a) $x \phi y = y/x$ d) $x \phi y = \dfrac{xy}{x + y}$

b) $x \phi y = x - y$

c) $x \phi y = x(x + y)$ e) $x \phi y = x^2 + xy^2 + y^4$?

This exercise really asks, "For which of the ϕ's defined on the right side of each equals sign can x and y be interchanged without altering the expression?" (The answer is d.)

In keeping with the emphasis on abstractions and structure the modern mathematics texts introduce the notions of group and field. A group is any collection of elements and an operation which satisfy several conditions. If we call the operation addition (though it may not be at all the addition of real numbers) then one condition is that the sum of any two elements must be another element of the collection. The associative property must apply to the operation of addition. There must be an element denoted by 0 such that $a + 0 = a$ for every element a of the collection. Finally, to each a of the collection there must be another element, denoted by $-a$, such that the sum of a and $-a$ is 0.

The simplest example of a group is the set of positive and negative whole numbers and the operation of addition. However, the importance of the group notion lies in the fact that there are many different collections of elements and an operation associated with each collection which form a group.

The concept of a field applies to a collection of elements and two operations, called addition and multiplication; each of these operations must possess the properties that the group operation possesses (with an exception with respect to 0) and the two operations are related by the distributive law, that is, $a \times (b + c) = a \times b + a \times c$.

The simplest example of a field is the collection of rational numbers (whole numbers and fractions) with the usual operations of addition and multiplication. As in the case of groups there are many collections of elements and operations that form a field.

The students are expected to learn not only the concepts of group and field but properties of these structures beyond the ones involved in the definitions. The study of these abstract structures and their properties is usually taught in the third or fourth year of the traditional undergraduate curriculum and only to majors in mathematics. In the new mathematics some of the concepts are taught in elementary school and the subject proper is taught in the fourth high school year.

Students are asked to learn abstract concepts in the expectation that if they learn these, the concrete realizations will be automatically understood. Thus if a student learns the general definition of a function, he will presumably understand the specific functions he will have to deal with; and if he learns what a field is, he will know all about the rational numbers, and other mathematical collections that form fields. Phrased in terms of abstract versus concrete, one can say that modern mathematics favors the abstract as the approach to the concrete.

That abstract subject matter was to be featured was indicated early in the movement. For example, in its 1959 report the Commission on Mathematics of the College Entrance Examination Board states (p. 20), "The goal of instruction in algebra should not be thought of exclusively or even largely as the development of manipulative skills. Rather instruction should be oriented toward the development and understanding of the properties of a number field." For the second semester of the twelfth year the Commission recommended a choice between introductory probability with statistical applications and an introduction to abstract algebra stressing groups and fields.

We have described the new contents of the modern mathematics curriculum. Presumably it meets the needs of young students entering into modern society. Apparently it should also appeal to them because, as the authors of *The Revolution in School Mathematics* put it

(p. 32): "He [the student] wants to know how mathematics fits into his world. And, happily, his world is full of fancy and abstractions. Thus students become interested in mathematics because it gives them quick access to a kind of intellectual adventure that is enticing and satisfying." Apparently human imagination is not dead.

Let us review the content of the new mathematics in the light of the claims made for the new curriculum.

The modern mathematics proponents have made much of the point that the mathematics taught in the traditional curriculum was all known before 1700 and that students were bored with such outdated mathematics. Moreover, the proponents claimed, the modern age requires totally new mathematics. How modern are the contents of modern mathematics?

Actually, most of the material in the modern mathematics curriculum is the traditional material. The old arithmetic, algebra, geometry, trigonometry, analytical geometry, and calculus are all there and are in fact the core of the new curriculum. Hence the word modern is certainly inappropriate. The charge that the traditional curriculum is outmoded is belied by the very admission of the modernists that they offer primarily a new approach to the old curriculum. Despite this fact the proponents began and continue their campaign on the major platform that the present age requires new mathematics for such applications as linear programming, operations research, game theory, quality control, and other fields. Actually these applications use traditional mathematics.

Even if the new curriculum had abandoned the older contents on the ground that it was all created by 1700 would the new curriculum be any better? Would the traditional curriculum then be three hundred years behind the times? Such an argument might apply to history but it has no force when applied to mathematics. Our subject is cumulative. The new builds on the old, and the old subject matter must be understood if the new developments are to be mastered. Hence a curriculum based solely on post-1700 mathematics would have no foundation.

One cannot defend all that is old. We have already noted (in Chapter 2) that the logarithmic solution of triangles, a favorite topic in the traditional trigonometry, has lost its importance. It can be discarded. But there is very little in the traditional curriculum that can be declared outmoded or useless today. Mathematics has been compared to a great tree ever putting forth new branches and new leaves but nevertheless having the same firm trunk of established knowledge. The trunk is essential to the support of the life of the entire tree.

However, the new curriculum does include some new content. In evaluating these new topics let us keep in mind that elementary and high school students do not know what they plan to do later in life. Even the few that think they do may change their minds several times. So whatever mathematics courses offer should be valuable in the full variety of careers that these students may take up.

The new topic that receives the most emphasis in modern mathematics is set theory. Now there is no question that the word "set" is useful. It means no more than collection, class, group, and the like, in the usual nontechnical sense. However, as we have already noted, students are asked to learn union and intersection of sets, subsets, the empty set, infinite sets, one-to-one correspondence between infinite sets, larger and smaller infinite sets and other concepts. All of this is a sheer waste of time. In very sophisticated and advanced theories of mathematics set theory plays a role, but in elementary mathematics it plays none. In fact it is almost certain that set theory was brought in to give the new mathematics the air of being sophisticated and advanced rather than because it is helpful. It happens to be one of the few topics of advanced mathematics that can be presented without requiring prohibitive background, and it is no doubt one of the few topics of advanced mathematics that some of the framers of the modern mathematics curricula could grasp.

Having introduced set theory the curriculum makers must use it. They invented a terminology and definitions which do so. Thus to speak of the "solution set of $x + 2 = 4$," instead of the value of x for which $x + 2 = 4$, uses set language. To speak of a triangle as "the union of three noncollinear points and the line segments joining them," replaces the expression three noncollinear points and the line segments joining them. The point of intersection of two lines is described as the set intersection of the two linear sets. Functions, as we noted earlier, are sets of ordered pairs. This definition of function is particularly unfortunate. What is important about a function, say, $y = x^2$, is that as x varies so does y and the values of y depend upon the values of x. All of this meaning is vitiated in the definition as a set of ordered pairs. In fact the set of pairs is infinite and cannot be ordered by the mind so as to see the important concept of variation. These uses of set theory, then, distort the basic concepts.

Professor Feynman, whom we have already cited as having examined textbooks for use in the state of California, says, "In almost all of the textbooks which discuss sets, the material about sets is never used—nor is any explanation given as to why the concept is of any particular interest or utility. The only thing that is said is 'the concept of sets is very familiar.' This is, in fact, true. The idea of sets is so familiar that I do not understand the need for the patient discussion of the subject over and over by several of the textbooks if they have no use for sets at the end at all."

The role of set theory in mathematics may be worth noting because it gives some indication of how the modern mathematics curriculum has approached mathematics. This role may be understood through an analogy. Suppose that, because of general dissatisfaction with our country's production of musicians, we decided to change music education. Some group of educators might come along and argue that we have failed to make progress in music because we are still teaching Bach, Beethoven, and Brahms. We must instead teach modern music. More than that, we must teach the physical foundation of music. Now the physical foundation of music is basically the physics of sounds, musical sounds in particular. Hence music students would be taught primarily the theory of sound to the detriment of playing music, listening to music, and appreciation of the great works of the past. One could of course maintain that the physics of musical sounds is a desirable subject in itself. This is true. But expertise in that subject, assuming even that it is pedagogically accessible to young people, will not produce musicians.

The same argument can be made with respect to the teaching of painting. One could teach the theory of colors and develop experts in this area. But these experts might not be able to paint a stroke—at least on the basis of their education in colors.

Similarly, set theory—though logically the foundation of a sophisticated and rigorous approach to mathematics —is of no use at all in understanding and learning to work with elementary mathematics.

As a matter of fact, set theory can be confusing even in the context where it is claimed to be most helpful, namely, in learning about numbers. The best the modern texts can say about the relation of number to set is that number is a property or a name of a set. This in itself is so vague as to be useless as a definition of a whole number. But the situation is worse than that. Given the two sets {1,2,3} and {3,4,5}, the union of the two sets is the set {1,2,3,4,5}, which contains only five objects. But adding the number of objects in the first set to the number of objects in the second gives 6. Likewise, the intersection of the two original sets is the set {3}. But the product of the two numbers represented by the two sets is 9. Hence union and intersection, the two basic operations with sets, do not correspond to adding and multiplying the numbers represented by the sets.

A critical examination of the uses of set theory in elementary and high school texts refutes the claim of the modernists that set theory unifies mathematics. Beyond using it artificially to define concepts, no significant use is made of the subject. The whole subject is in fact dropped and only the vocabulary survives in the later development. There are, of course, deep results within the area of set theory, but even the modernists recognize that these lie far beyond the province of elementary mathematics. In this matter we may compare set theory with elementary geometry. When first encountered, the axioms of geometry must seem immediately obvious to the student, and to this extent geometry and set theory start out on a par. But almost before he realizes that anything of moment is being developed in geometry, the student comes upon consequences that surprise and may excite him. From seemingly simple axioms and the deductive mode of reasoning emerge such unexpected and heady results as that the medians of a triangle meet in a point, as do the altitudes, the angle bisectors, and the perpendicular bisectors of the sides—and not just for one type of triangle but for every triangle. To a mathematically sensitive student, such results come as a never-to-be-forgotten revelation of the power of abstract mathematical reasoning. There is nothing comparable in the rudimentary treatment of set theory in the new mathematics curriculum.

Set theory is for elementary mathematics a hollow formalism which encumbers ideas that are far more easily understood intuitively. The attempt to involve it is almost ludicrous and a travesty of pedagogy. The theory of sets has not proved to be the elixir for mathematical pedagogy.

The emphasis on set theory has led to caustic criticisms. Not atypical is the following: "One writer urges that the students become active participants in an adventure in the learning of concepts. And what is this adventure? The students give the teacher their own examples of sets. They begin with sets of rather similar things; but soon the adventure has risen to the pitch where they can contemplate such sets as 'the nose of the notary, the moon, and the number 4.' The students have been led with great pedagogical skill to the breathtaking conclusion that any collection of things is a collection."

Another critic was equally severe. "Oh see, Johnny has a set of marbles. Look, look, Billy has a set of marbles. See Billy's set. Here comes Mary. Mary gets all the marbles. Mary gets the union of Johnny's set and Billy's set. See Mary's union."

The second new topic espoused by the new mathematics is bases of number systems. As we have already pointed out, this topic is not new historically nor is it new in the teaching of mathematics. It has been taught for generations in college algebra. Hence what is new is merely that the topic has been introduced at the elementary school level. The contention is that students will understand the usual base ten method of writing numbers better if they learn how to write them in any base.

Pedagogical problems are often difficult to resolve.

Perhaps a fair analogy to teaching bases early would be to teach students French while they are still learning the rudiments of English. Would they learn English better? It is more likely that learning two languages simultaneously would be confusing. One might improve and deepen his knowledge of English by learning French but this would best be done after the command of English is reasonably well established. Another analogy might be learning to play the piano and violin at the same time.

Another often encountered argument for the introduction of bases is that modern electronic computers use base two. Computers do use base two, but it would seem more pertinent to teach base two when students are about to learn how to use computers rather than in the arithmetic of the second, third, or fourth grade.

Congruence is a third new topic. This, we may recall, involves arithmetic modulo twelve, or five, or six. There is no application of this concept on an elementary level. It may intrigue students but at the expense of learning something more important and perhaps equally intriguing. There seems to be nothing to recommend this topic for the elementary level except its novelty.

As to inequalities—another of the topics now featured in modern mathematics texts—this, like bases, was commonly taught at the college algebra level and all that has been done is to bring this topic down to the high school level. Very little can be done with this topic at the high school stage. Even when taught at the college level it was not used for some time thereafter. Hence there is even less point to teaching it at the lower level.

Boolean algebra, another of the new topics, is very much like set theory so far as union and intersection are concerned. Primarily it is a step in the direction of mathematical logic. Teaching it at the high school level is now defended on the ground that it is used in the design of switching circuits, especially in computers. The defense seems weak in view of the fact that elementary and high school students do not know what careers they will follow, so that elementary and high school education should be broad rather than pre-professional. How many high school students will design switching circuits later in life?

Still another topic of the modern curriculum is symbolic logic. There are many reasons for not teaching it to young people. There seems to be a prevalent but nonetheless mistaken belief that symbols explain concepts or ideas. Presumably the meaning of a joint assertion of two propositions such as "It is raining and I am going for a walk" is explained and even defined by a truth table. This is putting the cart before the horse. How was the truth table obtained? Evidently we have to know in advance that the joint assertion of two propositions is true if and only if both are true. Then we can construct the truth table. The attempt to reason through the meanings supplied by truth tables is inefficient, clumsy and even baffling. No mathematician or lawyer reasons in this way. And no mathematician except a specialist in symbolic logic uses symbolic logic. Rather, every mathematician thinks intuitively and then presents his arguments in a deductive form using words, familiar mathematical symbols and common logic. Specialists in foundation problems, who have to worry about the ambiguities and imprecision of ordinary language, do resort to symbolic logic. But even these people know intuitively what they want to say and *then* express their thoughts in special symbols. The symbolic logic does not control or direct the thinking; it is merely the compact written expression of the real thinking. Indeed, one must make sure that the symbols express what is intended rather than that the symbols tell him what he means to say. Not only is symbolic logic not used by most mathematicians but those who do use it do their effective thinking in common language.

One of the subjects recommended for the twelfth-grade course and rather widely adopted is an introduction to abstract algebra. In many texts this course stresses matrices, the nature of which we described earlier. There is an extensive algebra of matrices and students are taught how to add, subtract, multiply and divide matrices as well as many other operations. However, matrices are themselves abbreviated forms of what are called transformations and the students at the twelfth-grade level have no background in the significance and uses of transformations. Hence they learn merely to manipulate matrices to no purpose, and they cannot have any feeling for or make sense of what is accomplished because they do not know the context in which matrices are useful. This manipulation of matrices becomes, then, a series of mechanical tasks—as mechanical as the traditional teaching of algebra, which has been justly criticized. The algebra of ordinary numbers is at least a necessary step to progress in elementary mathematics; and while no case can be made for meaningless mathematics of any sort, there is some reason to teach the algebra of ordinary numbers to high school students. There is no justification for teaching matrices at the twelfth-grade level.

Another component of the recommended abstract algebra course is the study of groups and fields. We have already explained that these are abstract formulations of various, more concrete algebras. The abstract versions are studies of structures common to the concrete cases. The argument against teaching such structures is simply that it is premature. One might as well try to teach the structure of all languages—they do have common features—to a child who has yet to learn the English language. Once a student has learned the algebras of real numbers, complex numbers, rational functions, matrices, vectors, transformations, and congruences, it may enlighten him to know that these algebras do or do not have some common features. One can then even prove a theorem about groups and the theorem will apply to all those concrete algebras which form groups. There is, in other words, an advantage to proving the abstract theorem once and for all so as to cover many special cases in one swoop. However, the abstractions themselves are meaningless if not preceded by effective learning of specific algebras, and the advantage of proving a general theorem in the abstract algebra will hardly be impressive to one who knows only one instance of a group or a field—and knows it imperfectly at that.

The very point of an abstract formulation is that it unifies and reveals common properties in concrete and familiar branches of mathematics. Therefore, abstraction is not the first stage but the last stage in a mathematical development. It may give insight, but only into concrete structures already well learned. It unifies, but only what one already knows. Without much previous knowledge of concrete cases the abstract concepts remain empty, arbitrary children of mathematical fantasy. To confront youngsters with abstractions that lie above their level of maturity is to create bewilderment and revulsion rather than increased knowledge. In brief, the highly abstract concepts cannot be exploited at an elementary level.

There is another objection to teaching the abstract structure prematurely. It is true that the rational numbers, the real numbers, and the complex numbers each have the properties of a field. That is, the sum, product, difference and quotient of any two rational numbers is a rational number and the same is true for real numbers and for complex numbers. One could therefore argue that the students do have three concrete examples of a field. Should the field concept then be taught early? The answer is still no, because the field properties are those which are *common* to all these systems of numbers, so they automatically wipe out any distinguishing features. But the operations with rational numbers are different from those for real numbers and the latter are different from those for complex numbers. Thus the product of the fractions 3/4 and 7/8 is 21/32; the product of $\sqrt{2}$ and $\sqrt{3}$ is $\sqrt{6}$; and the product of $2 + 3i$ and $4 + 5i$ is $-7 + 22i$. Until the student can at least operate freely with rational and real numbers, there is little point in his knowing that $a \times b = b \times a$ is a property of fields. A student who knew perfectly the properties of a field could not necessarily make change in a grocery store, much less balance a check book. Let us keep in mind that the more general the mathematical concept, the emptier it is.

To put it another way, fields do not explain the types of numbers and operations with them. In fact it is just the reverse. A good understanding of the several number systems explains the concept of a field. Hence, the familiar argument that it is efficient to teach the abstract concept early because it comprises several concrete cases at once is groundless. So far as efficiency is concerned, the time that is wasted is the time spent teaching the abstract concept. When one takes into account that perhaps fifty per cent of the students entering college cannot add and multiply fractions, especially if letters are involved, one can see where the emphasis must be put.

Psychologically the teaching of abstractions first is all wrong. Indeed, a thorough understanding of the concrete must precede the abstract. Abstract concepts are meaningless unless one has many and diverse concrete interpretations well in mind. Premature abstractions fall on deaf ears.

In a real sense one cannot teach an abstraction. The

difficulty posed to the student is analogous to giving him a correct biological definition of dogs and then showing him a poodle and a collie as examples. When presented with a bull terrier and asked if that is a dog, the student may still be baffled. The biological definition contains so many broad and technical terms that he may not really understand it; if so he cannot apply it.

Abstractions must grow on people. They cannot be handed down *ex cathedra.* As people's experiences with the different varieties of functional relationships and their peculiarities increase, they will gradually see the wealth of situations which have a basic idea in common; at this stage they will be enlightened if this common idea is pointed out. They will also appreciate the various conditions or qualifications that must be incorporated in a general definition.

To start with a general concept and then to use it only in special cases, which is the practice in modern mathematics, is pedagogically absurd for another reason. One may wish to teach a six-year-old a little about dogs so that he may play with them and at the same time be cautious around them. Would one start with the biological definition? Leaving aside whether the definition is meaningful to the youngster, we should ask whether it is helpful to know it. Would it not be far more practical for this youngster to learn about the particular dogs that he will meet in his environment? So it is with the mathematical concepts. The general definition of a function is useless as long as the immediate objective is to learn about $y = 3x$, $y = 3x + 7$, $y = x^2$ and the like. In fact, the general definition burdens the student with a mystery which clouds all his subsequent thinking. A year or two after learning a definition of function applicable only to simple cases, he may need a slightly broader notion of function—for example, when he meets functions of two or three variables. Even then the general definition is useless. It is sufficient to extend the term function to include, for example, $z = x^2 + y^2$.

The human mind operates no differently in mathematics than in political or social thought. One may preach the brotherhood of man, but this preaching does not get people to understand and practice brotherhood. Teaching young children to get along with and respect others of different races and creeds in daily acts may achieve brotherhood, but the general ethical principle will make no impression. Students who are taught abstractions before they have acquired the rich experience which in fact led to those abstractions may acquire superficial knowledge and be able to mouth words. But they cannot be said to really understand these abstractions.

The modern mathematician's preference for abstractions is reminiscent of a story. The principal of a school boasted of how much he liked his students and about how much he was concerned for their welfare. One day he saw a student walking on a freshly laid concrete sidewalk. He rushed over and roughly yanked the student off the fresh concrete. A teacher who saw the incident reminded the principal of his professed concern for students and chided him for his rough handling of the boy. The principal replied, "I like students in the abstract but not in the concrete."

The new mathematics advocates, countering the charge of too much abstraction, have cited the Harvard psychologist Jerome S. Bruner who said, "Any subject can be taught in some intellectually honest form to any child at any stage of development." The saving feature of this doctrine is its vagueness. Leaving aside the question of whether the particular abstractions that any one group may be interested in promoting warrant priority, one wonders how one could present the substance of Kant's *Critique of Pure Reason* even to high school students. What can happen and does happen when Bruner's doctrine is taken too seriously is that students accept the abstractions docilely and are as understanding and as critical of what they are taught as children are when they learn a catechism.

The total picture presented by the new topics in the modern curriculum is hardly impressive. Some items, such as bases and inequalities, were merely pushed down from the levels at which they were previously taught, with no gain and much reason to question. Others, such as congruences, are at the elementary level pointless novelties. Still others, such as Boolean algebra and symbolic logic, are specialties that should be reserved for specialists. And finally, the advanced topics, set theory, matrices, and abstract algebra, seem to have been deliberately chosen to show that the curriculum has caught up with advances in mathematics—even though these advances have no proper place or function in the training of youngsters. So far as the advanced topics are concerned, the "reformers" seem to be under the impression that what smacks of modern mathematics is modern mathematics—rather like the small boy who thinks that because he wears long pants he is a grown-up.

Though several capable and broadly educated mathematicians participated in framing the numerous versions of the new mathematics, their contributions were highly diluted. The new mathematics as a whole is a presentation from the point of view of the shallow mathematician, who can appreciate only the petty deductive details and minor pedantic, sterile distinctions such as between number and numeral and who seeks to enhance trivia with imposing-sounding terminology and symbolism. This mathematics offers an abstract, rigorous version that conceals the rich and fruitful essence and emphasizes uninspiring generalities isolated from all other bodies of knowledge. It stresses final sophisticated versions of simple ideas while treating superficially the deeper ones—

and so necessarily assumes a dogmatic character. The formalism of this curriculum can lead only to an erosion of the vitality of mathematics and to authoritarianism in teaching, the rote learning of new routines far more useless than the traditional routines. In brief, it presents form at the expense of substance and substance without pedagogy.

Perhaps the decisive criticism of the modern mathematics program was made unwittingly by a professor who was evidently pleased with it and who intended his remark to be words of praise: "If they [the students] are going to fail mathematics, they might just as well fail good mathematics."

When one criticizes the modern mathematics program one often finds that the listener, wishing to be sympathetic and not knowing where to stand, remarks that undoubtedly the truth lies between the traditional and the modern mathematics versions. Such a compromise may be correct in some controversies, but it is not applicable here. If someone argues that the earth turns from east to west and another that the earth turns from west to east, one cannot compromise, however well-meaning one may be, by saying that the truth lies between these alternatives. As we shall see later in Chapter 11, the proper reform is diametrically opposite to the path taken by modern mathematics and lies, so to speak, on the other "side" of traditional mathematics.

9

The Testimony of Tests

"And lo! Ben Adhem's name led all the rest."

Leigh Hunt

Despite all of the seemingly valid criticisms of both the traditional and the modern mathematics curricula, one feels that there should be some more objective criteria for determining whether a given program is more suitable or better than another. One would think that the framers of the modern mathematics curriculum would have experimented with many groups of children and teachers and thus produced some evidence in favor of their programs before urging them upon the country. The sad fact is that most of the groups undertook almost no experimental work. The sole significant exception was the University of Illinois Committee on School Mathematics headed by Professor Max Beberman. And even this group never offered any evidence for the superiority of its curriculum.

Beberman started in 1952 and for the first few years experimented with classes in the University of Illinois high school. He also trained teachers to teach the new material in other schools and was prepared to spend

124

some years in testing his material. But by 1955 the Commission on Mathematics entered the picture and began to proclaim what it advocated in its final report of 1959. In 1958 The School Mathematics Study Group (SMSG) was organized. During the summer of 1958 fourteen experimental topics for grades seven and eight were written. These were tried by about one hundred teachers in twelve centers during the school year 1958–1959. Some revisions in these units were made during the summer of 1959. By this time the SMSG writing groups had completed texts for all of the grades seven through twelve. These were used experimentally during the 1959–1960 school year and revised during the summer of 1960. Without further ado the SMSG began to sell the curriculum to the country. Seeing these competitors and others ready to market their products, Professor Beberman probably felt that his work would be lost in the shuffle and he began to campaign actively for his version of the modern mathematics curriculum. Thereafter, experimental work practically vanished and the rush to secure leadership took precedence over all other activity.

During the fall of 1960 the National Council of Teachers of Mathematics, which had put together its own curriculum through its Secondary School Curriculum Committee, conducted eight regional conferences in various parts of the United States. The purpose of these conferences was to give school administrators and mathematics supervisors information that would enable them to institute one of these "new and improved" mathematics programs. In other words, the millennium had arrived and the country was invited to rejoice and partake in it. The pamphlet *The Revolution in School Mathematics,* published in 1961, reported what was discussed during the 1960 conferences and states in its Preface, "We are now in a position to make a concerted effort toward rapid improvement of school mathematics. The general pattern is clear and the necessary materials of instruction are at hand. . . . We regard each of the new programs as a sample of an improved curriculum in mathematics which deserves the consideration of those interested in devising better mathematics programs." Thus by 1960— after efforts had been devoted almost entirely to writing subject matter, while testing was practically ignored— the new curricula were being sold to the country.

Professor Edwin E. Moise, who participated in the SMSG writing effort and gave talks in favor of that modern mathematics curriculum, unwittingly testified to the lack of experimentation. In his article in the booklet *Five Views of the New Math* (published by the Council for Basic Education) Moise said, "One thing was obvious, however, as soon as the books were written, and before they were tried: The improvement in intellectual content was so great that they surely would produce either an educational improvement or a collapse of classroom morale."

What kind of objective evidence should one look for to establish the superiority of any curriculum? One thinks almost immediately of tests. But one must be wary of the significance of mathematics tests. Presumably the mathematics courses have taught students to think through problems, to discover results for themselves, and to acquire insight into the concepts and proofs they have learned. Actually tests do not and to a large extent cannot measure such values. The usual tests require that a student answer a fair number of questions in a limited amount of time. If a question really called for thinking through a new type of problem or discovering a new result it would require so much time that average students would not be able to do well in the time allowed. Even if the student made intelligent and significant attempts to answer such questions his failure to reach a positive result would probably mean that he would receive little or no credit.

Hence, mathematics tests usually and almost perforce call for handing back information that has been learned and is merely being reproduced. The ability to memorize is the major faculty that is actually tested. While the acquisition of information is one objective of mathematics education, it is not supposed to be the sole or most important goal. Though most teachers would deny that they are testing memorization, their behavior belies their words. In Chapter 2 we noted that teachers do not permit their students to use their books during tests. But if the tests call for thinking on the part of the student, what vitiation of the tests would result from the use of books?

Let us assume, however, that to secure objective evidence rather than the opinions of teachers we must resort to tests, and let us further assume that the tests do give evidence of learning. We must still be wary of what the results mean, particularly in the case of the modern mathematics movement.

In most schools the modern mathematics curriculum is offered to the better or best students. In fact many of the proponents of modern mathematics now state explicitly that this program is intended for the college-bound and college-capable students. Clearly these students will do better than the average.

It is also true that a great many teachers of modern mathematics courses are motivated to do a better job. There are several reasons. It is the more able and more enterprising who want to try new material. Some receive extra pay or other benefits for experimenting or teaching other teachers how to teach modern mathematics. Still others have been convinced by the literature advocating modern mathematics that it is a superior curriculum and respond to the challenge of presenting it. Finally, many teachers participated in fashioning one or

another version of the modern curriculum and are determined to show that it is superior. When in addition these
teachers tell the students that they are specially chosen
and that they are participating in a noble experiment, the
students usually respond to such compliments and put
forth extra efforts. Certainly students taught under any
of the preceding conditions will do better. Incidentally,
the effect achieved by making students believe that they
are the key people in an important experiment is known
as the Hawthorne effect.

Tests have been given to small groups by teachers and
curriculum designers who are proponents of modern mathematics. The evaluations usually claim that students were
able to learn the new material and yet do as well in the
techniques of traditional mathematics as students who
are taught only the latter. This class of evaluations is suspect. Beyond the reasons previously cited for questioning
the results of tests are other factors. The tests are not
standardized; hence it is very easy for the teacher to bias
the questions so that the knowledge of the traditional
mathematics that is called for is minimal. Moreover, the
results of any test must be interpreted. Let us suppose
that the questions testing traditional material were ones
used to test one hundred thousand students whose average grade was sixty per cent. The group being tested by
the individual teacher might make an average grade of
seventy per cent on the traditional material. Has this
group done better? Not necessarily. This small group
may be among the better students and might have aver-
aged seventy per cent in the test given to the one hundred
thousand students. Moreover, if this group were taught
only traditional mathematics it might have attained an
average grade of ninety per cent on the traditional material. Could not the general intelligence of this small
group be tested so that one would know how it compares
with that of the one hundred thousand students? Any at-

tempt to do this raises the entire issue of the reliability of intelligence tests. We know very little about what intelligence is and how to test for it.

The picture of what is accomplished in modern mathematics courses is confused by another factor. A number of texts and courses based on them contain primarily traditional material which is dressed up (contaminated?) with a smattering of modern mathematics. Chapters on modern mathematics topics are interspersed with chapters on traditional mathematics with no integration of the two approaches. Incidentally many of these hybrid texts (one could use a more apt word to describe the progeny of this unholy wedlock of traditional and modern mathematics) are the output of hypocritical authors who evidently wish to capitalize on both markets, modern and traditional. As a matter of fact, some of the strongest advocates of the new mathematics have written such texts. Other texts begin with a chapter on set theory, then turn to traditional mathematics and thereafter never refer to set theory or any other topic of modern mathematics. Still other texts go beyond the introductory chapter on set theory only to the extent of using modern mathematics terminology in the body of the text, but these books still teach primarily traditional mathematics. Even totally traditional texts are now titled modern, new or contemporary.

When the teachers of courses based on such texts are asked whether they are teaching modern mathematics they will usually reply affirmatively. They are under pressure from chairmen, principals and superintendents to be up-to-date and since this means modern mathematics, they profess to be teaching it. If their students do well in tests based on these courses, the impression given is that students can and do learn modern mathematics, when in fact they are being taught and tested on the traditional mathematics.

One might be inclined to believe that these traditional texts "souped up" with a bit of modern mathematics could not be in wide use and so could not affect the evaluation of modern mathematics. Actually, such texts are the most popular ones because they cater to the traditionally oriented teacher who wishes or is obliged to claim that he is teaching the new mathematics. He can pretend to do so by actually presenting the minimal amount contained in the texts, or he can omit this material without affecting the treatment of the traditional subject matter because the two classes of material are not integrated.

One international test known as the International Study of Achievement in Mathematics may be worth mentioning because it was conducted in 1964, by which time many of the United States entries had had some modern mathematics. The test actually consisted of several tests given to students of different age levels. At all levels the United States ranked low and this was especially true of the thirteen-year-old group, which ranked tenth. Japan, incidentally, ranked first at all the age levels and its students had only traditional mathematics. Despite the poor showing of our students, some proponents of modern mathematics tried to interpret the statistics as showing that those students who took modern mathematics did better than those who took only traditional mathematics. But the evidence was meager and dubious. Though the results might seem to favor the teaching of traditional mathematics even this conclusion may be erroneous. The countries differ vastly in their emphasis on mathematics. For example, the British high school student who takes up mathematics practically specializes in the subject. Moreover, the poorer students never get to an academic high school; they are eliminated by tests given at the age of eleven and are sent to vocational schools. In the Soviet Union and Japan the young boys and girls must do exceedingly well to gain admission to any college and they work hard to excel. Students in the United States are not under such pressure.

What one wishes to test, to the extent that tests are significant, are normal students taught under normal conditions, and for these we have no results. In fact no large-scale testing of the quality of the modern mathematics program has been undertaken. At present the amount of effort devoted to assessing properly the claims of the proponents of modern mathematics is negligible in comparison with the claims. The superior understanding which the modern mathematics approach is supposed to provide has not been demonstrated by tests or by any other objective measures.

The education of students in the modern mathematics curriculum has been in vogue long enough so that students have entered college with this background. Do these students perform better because they received a possibly superior education? It is almost impossible to answer this question. No large-scale tests of these students have been made. Moreover, it is difficult to segregate those who have had modern mathematics because, as we have already indicated, many courses pretending to be modern are really either mixtures of traditional and modern or include just a smattering of modern mathematics. An informal consensus of college teachers is that students are now weaker in technique than those of ten or more years ago. But this fact, if it is a fact, does not necessarily point to defects in the modern mathematics curriculum. The pressure on young people to secure a college education has brought many more students to college who are not as well prepared in all subjects and who are less motivated. Also, mathematics education is being speeded up at a time when slowing down would seem wiser. High school students used to take a full course in synthetic plane geometry. Under the modern mathematics program part of this course is devoted to a modicum of analytic and solid geometry. Previously, either in the last year of high school or first year of college, students took advanced algebra and solid geometry. This material has been practically eliminated. Further, students used to take a full semester of analytic geometry before taking the calculus. The analytics course, beyond introducing the major idea of relating equations and curves, also enabled students to improve their algebra, geometry and trigonometry. For the last ten years analytic geometry has been submerged in the calculus and very little of it is taught. Thus the student enters calculus with far less preparation than he used to have and is weaker as a consequence.

There are, however, other indications, apart from tests, that all is not well with modern mathematics. In a speech given at a Symposium sponsored by the Thomas Alva Edison Foundation and held in Pittsburgh in November 1960, Professor Beberman confessed that he was wrong in putting rigor into geometry. He even exclaimed that he could not understand how he could have made that mistake.

At this same Symposium, Professor Edward G. Begle, the director of the largest and most influential curriculum group, the School Mathematics Study Group, said at the very outset of his speech, "In our work on curriculum we did not consider the pedagogy."

Somewhat later in a speech given at a University Symposium on Mathematics, held at Ohio State University on November 16, 1962, Professor Beberman cast further doubts on the wisdom of his program, then ten years old. "I think in some cases we have tried to answer questions that children never raise and to resolve doubts they never had, but in effect we have answered our own questions and resolved our own doubts as adults and teachers, but these were not the doubts and questions of the children."

From this he progressed with commendable honesty to outright criticism. In another speech given on December 30, 1964, at the 1964 Christmas meeting of the National Council of Teachers of Mathematics held in Montreal, Dr. Beberman confessed that "we're in danger of raising a generation of kids who can't do computational arithmetic." He admitted that the new curriculum had failed to relate mathematics to the real world and that pedagogical principles had been ignored. Because excessive emphasis was being placed on esoteric branches of mathematics at the expense of fundamentals, and because of the hasty introduction of the new mathematics in the elementary schools, Professor Beberman feared that "a major national scandal" may be in the making.

It is clear from these and many other indications that Professor Beberman had not been satisfied with the work of his group, and in the mid-winter of 1971–72 he went to England to study some new experimental curricula being fashioned there. Unfortunately Professor Beberman died shortly after arriving in England. There is some question as to what the Illinois group will do without his dynamic leadership.

The clearest evidence that the modern curriculum as expounded in the early 1960s is not satisfactory is found in a number of other statements by Professor Begle. He has admitted in numerous speeches that the SMSG curriculum has minimized the acquisition of skills and has failed to present the relationship of mathematics to allied subjects. He then announced in an open letter (which appeared in *The Mathematics Teacher* for April 1966 and in *Science* for February 1966) the intention to devise a totally new curriculum for grades seven to twelve. In this letter he states that the SMSG Advisory Board "feels that longer-range planning and experimentation is necessary and should be started now. This must be done to prevent the present materials from becoming frozen into a new orthodox pattern that would require another upheaval a few years from now."

But if the work from 1958 to 1966 was really good, why should there be any concern about the curriculum becoming frozen or about the necessity for a new upheaval? The nearest thing to an answer one can find in Professor Begle's letter is that the new curriculum "will be responsive to the rapidly developing needs for mathematics in our society."

A committee was appointed in 1966 to make plans for the writing of the new curriculum. By 1972 the revision concerned with the junior high school was completed.

A special version for low achievers and a tenth-grade course which would make the transition to the older SMSG eleventh- and twelfth-grade materials were also fashioned. In the SMSG *Newsletter* of February 1972, Professor Begle described these programs. Though the content of the new junior high curriculum varies somewhat from the previous material, the chief features are not essentially different. For example, ". . . structure is still definitely one of the unifying themes." Moreover, the elementary-school curriculum and the older material for grades ten, eleven and twelve were left untouched.

The program projected in 1966 will not be completed. It is known in professional circles that neither Professor Begle nor his sponsors have been satisfied with the partial revision. The financial support has been withdrawn and SMSG is being disbanded.

Beyond the evidence just presented and the criticisms made in the preceding chapters, the arguments against the new mathematics are supported by the judgments of men who, we have every reason to believe, are impartial.

Numerous critical articles have appeared in such professional journals as *The Mathematics Teacher*. There is no doubt that many more teachers, who would like to express their disapproval of modern mathematics, fear to do so because it might incur the displeasure of their chairmen, principals or superintendents.

One expression of criticism should be noted. A few college professors (this author was one of them) got together, drafted a protest against the movement, and asked mathematicians if they wished to endorse it. It would have been possible to get hundreds of signatures, but since the objective was just to show significant opposition, it was decided that about seventy-five active, mature mathematicians should be solicited. The memorandum, entitled "On the Mathematics Curriculum of the High School," was published in *The Mathematics Teacher* of March 1962 and in the *American Mathematical Monthly* of March 1962. It warrants perusal and is reproduced here.

ON THE MATHEMATICS CURRICULM OF THE HIGH SCHOOL

The following memorandum was composed by several of the undersigned and sent to 75 mathematicians in the United States and Canada. No attempt was made to amass a large number of signatures by canvassing the entire mathematical community. Rather, the objective was to obtain a modest number from men with mathematical competence, background, and experience and from various geographical locations. A few of the undersigned, whose support is indeed welcomed, volunteered their names when they learned about the memorandum from a colleague.

The mathematicians of this country now have a more favorable climate in which to develop and gain acceptance of improvements in mathematics education. Indeed a number of groups have recognized the opportunity and are working hard and with the best of intentions to utilize it.

It would, however, be a tragedy if the curriculum reform

should be misdirected and the golden opportunity wasted
There are, unfortunately, factors and forces in the current
scene which may lead us astray. Mathematicians, reacting
to the dominance of education by professional educators
who may have stressed pedagogy at the expense of content,
may now stress content at the expense of pedagogy and be
equally ineffective. Mathematicians may unconsciously
assume that all young people should like what present day
mathematicians like or that the only students worth culti-
vating are those who might become professional mathematicians. The need to learn much more mathematics today
than in the past may cause us to seek shortcuts which,
however, could do more harm than good.

In view of the possible pitfalls it may be helpful to
formulate what appear to us to be fundamental principles
and practical guidelines.

1. For whom. The mathematics curriculum of the high
school should provide for the needs of all students: it
should contribute to the cultural background of the general student and offer professional preparation to the
future users of mathematics, that is, engineers and scientists, taking into account both the physical sciences which
are the basis of our technological civilization, and the social sciences which may need progressively more mathematics in the future. While providing for the other students
the curriculum can also offer the most essential materials
to the future mathematicians. Yet to offer such subjects to
all students as could interest only the small minority of
prospective mathematicians is wasteful and amounts to
ignoring the needs of the scientific community and of
society as a whole.

2. Knowing is doing. In mathematics, knowledge of any
value is never possession of information, but "know-how."
To know mathematics means to be able to do mathematics:
to use mathematical language with some fluency, to do
problems, to criticize arguments, to find proofs and, what
may be the most important activity, to recognize a mathematical concept in, or to extract it from, a given concrete
situation.

Therefore, to introduce new concepts without a sufficient background of concrete facts, to introduce unifying concepts where there is no experience to unify, or to harp on the introduced concepts without concrete applications which would challenge the students, is worse than useless: premature formalization may lead to sterility; premature introduction of abstractions meets resistance especially from critical minds who, before accepting an abstraction, wish to know why it is relevant and how it could be used.

3. Mathematics and science. In its cultural significance as well as in its practical use, mathematics is linked to the other sciences and the other sciences are linked to mathematics, which is their language and their essential instrument. Mathematics separated from the other sciences loses one of its most important sources of interest and motivation.

4. Inductive approach and formal proofs. Mathematical thinking is not just deductive reasoning; it does not consist merely in formal proofs. The mental processes which suggest what to prove and how to prove it are as much a part of mathematical thinking as the proof that eventually results from them. Extracting the appropriate concept from a concrete situation, generalizing from observed cases, inductive arguments, arguments by analogy, and intuitive grounds for an emerging conjecture are mathematical modes of thinking. Indeed, without some experience with such "informal" thought processes the student cannot understand the true role of formal, rigorous proof which was so well described by Hadamard: "The object of mathematical rigor is to sanction and legitimize the conquests of intuition, and there never was any other object for it."

There are several levels of rigor. The student should learn to appreciate, to find and to criticize proofs on the level corresponding to his experience and background. If pushed prematurely to a too formal level he may get discouraged and disgusted. Moreover the feeling for rigor can be much better learned from examples wherein the proof settles genuine difficulties than from hair-splitting or endless harping on trivialities.

5. Genetic method. "It is of great advantage to th
student of any subject to read the original memoirs on tha
subject, for science is always most completely assimilate
when it is in the ascent state," wrote James Clerk Max
well. There were some inspired teachers, such as Erns
Mach, who in order to explain an idea referred to its ger
esis and retraced the historical formation of the idea. Thi
may suggest a general principle: The best way to guide th
mental development of the individual is to let him retrac
the mental development of the race—retrace its great lines
of course, and not the thousand errors of detail.

This genetic principle may safeguard us from a commor
confusion: If A is logically prior to B in a certain system
B may still justifiably precede A in teaching, especially i
B has preceded A in history. On the whole, we may ex
pect greater success by following suggestions from th
genetic principle than from the purely formal approach t
mathematics.

6. "Traditional" mathematics. The teaching of mathe
matics in the elementary and secondary schools lags fa
behind present day requirements and highly needs essentia
improvement: we emphatically subscribe to this almos
universally accepted opinion. Yet the often heard assertior
that the subject matter taught in the secondary schools i
obsolete should be closely scrutinized and should not be
taken simply at face value. Elementary algebra, plane and
solid geometry, trigonometry, analytic geometry and the
calculus are still fundamental, as they were fifty or a hun
dred years ago: future users of mathematics must learn al
these subjects whether they are preparing to become math
ematicians, physical scientists, social scientists or engi
neers, and all these subjects can offer cultural values to the
general students. The traditional high school curriculun
comprises all these subjects, except calculus, to some ex
tent; to drop any one of them would be disastrous.

What is bad in the present high school curriculum is not
so much the subject matter presented as the isolation of
mathematics from other domains of knowledge and in
quiry, especially from the physical sciences, and the isola-

tion of the various subjects offered from each other; even the techniques and theorems within the same subject appear as isolated, disconnected tricks to the student, who is left in the dark about the origin and the purpose of the manipulations and facts that he is supposed to learn by rote. And so, unfortunately, it often happens that the material offered appears as useless and boring, except, perhaps, to the few prospective mathematicians who may persist despite the curriculum.

7. "Modern" mathematics. In view of the lack of connection between the various parts of the present curricula, the groups working on new curricula may be well advised in seeking to introduce unifying general concepts. We think, too, that judicious use of sets and of the language and concepts of *abstract* algebra may bring more coherence and unity into the high school curriculum. Yet, the spirit of modern mathematics cannot be taught by merely repeating its terminology. Consistently with our principles, we wish that the introduction of new terms and concepts should be preceded by sufficient *concrete* preparation and followed by genuine, challenging application and not by thin and pointless material: one must motivate and apply a new concept if one wishes to convince an intelligent youngster that the concept warrants attention.

We cannot enter here into detailed analysis of the proposed new curricula, but we cannot leave unsaid that, in judging them on the basis of the guidelines stated above (Sections 1–5), we find points with which we cannot agree.

Of course, not all mathematicians have the same taste. Mathematics has many aspects. It can be regarded as an instrument to understand the world around us: mathematics presumably possessed this value for Archimedes and Newton. Mathematics can also be regarded as a game with arbitrary rules where the principal consideration is to stick to the rules of the game: some such view may be considered suitable for certain problems of foundations. There are several other aspects of mathematics, and a professional mathematician may favor any one. Yet when it comes to teaching, the choice is not a mere matter of taste.

We may expect that an intelligent youngster would war to explore the world around him, but we cannot expec him to learn arbitrary rules: why just these and not others

At any rate, we fervently wish much success to the workers on the new curricula. We wish especially that the new curricula should reflect more the connection between mathematics and science and carefully heed the distinction between matters logically prior and matters which should have priority in teaching. Only in this way can we hope that the basic values of mathematics, its real meaning, purpose, and usefulness will be made accessible to all students, including of course, the prospective mathematicians. The recently expressed "widespread concern about a trend to excessive emphasis on abstraction in the teaching of mathematics to engineers" * points in the same direction.

Lars V. Ahlfors, Harvard University

Harold M. Bacon, Stanford University

Clifford Bell, University of California, Los Angeles

Richard E. Bellman, Rand Corporation

Lipman Bers, New York University

Garrett Birkhoff, Harvard University

R. P. Boas, Northwestern University

Alfred T. Brauer, University of North Carolina

Jack R. Britton, University of Colorado

R. C. Buck, University of Wisconsin

George F. Carrier, Harvard University

Hirsh Cohen, IBM

Richard Courant, New York University

H. S. M. Coxeter, University of Toronto

Dan T. Dawson, Stanford University

Avron Douglis, University of Maryland

Arthur Erdelyi, California Inst. of Technology

Walter Freiberger, Brown University

K. O. Friedrichs, New York University

Paul R. Garabedian, New York University

David Gilbarg, Stanford University

Sydney Goldstein, Harvard University

* First Summer Study Group in Theoretical and Applied Mechanics Curricula, Boulder, Colorado, June 1961.

Herman Goldstine, International Business Machines Corp.

Herbert Greenberg, International Business Machines Corp.

John D. Hancock, Alameda State College

Charles A. Hutchinson, University of Colorado

Mark Kac, Rockefeller Institute

Wilfred Kaplan, University of Michigan

Aubrey J. Kempner, University of Colorado

Lucien B. Kinney, Stanford University

Morris Kline, New York University

Ignace I. Kolodner, University of New Mexico

Rudolph E. Langer, University of Wisconsin

C. M. Larsen, San Jose State College

Peter D. Lax, New York University

Walter Leighton, Western Reserve University

Norman Levison, Massachusetts Institute of Technology

Hans Lewy, University of California, Berkeley

W. Robert Mann, University of North Carolina

M. H. Martin, University of Maryland

Deane Montgomery, Institute for Advanced Study

Marston Morse, Institute for Advanced Study

Zeev Nehari, Carnegie Institute of Technology

Jerzy Neyman, University of California, Berkeley

Frederick V. Pohle, Adelphi College

H. O. Pollak, Bell Telephone Laboratories

George Pólya, Stanford University

Hillel Poritsky, General Electric Co.

William Prager, Brown University

Murray H. Protter, University of California, Berkeley

Tibor Rado, Ohio State University

Warwick W. Sawyer, Wesleyan University

Max M. Schiffer, Stanford University

James B. Serrin, University of Minnesota

Lehi T. Smith, Arizona State University

I. S. Sokolnikoff, University of California, Los Angeles

Eli Sternberg, Brown University

J. J. Stoker, New York University

A. H. Taub, University of Illinois

Clifford E. Truesdell, Johns Hopkins University

R. J. Walker, Institute for Defense Analyses and Cornell University

Wolfgang Wasow, University of Wisconsin

André Weil, Institute for Advanced Study

Alexander Wittenberg, Laval University

10

The Deeper Reasons for the New Mathematics

"A modern mathematician would prefer the positive characterization of his subject as the study of general abstract systems, each one of which is built of specified abstract elements and structured by the presence of arbitrary but unambiguously specified relations among them."

Marshall H. Stone

In view of the defects in the modern mathematics program and the failure to remedy the defects of the traditional curriculum, why was the modern mathematics program devised and promoted? Moreover, since the claimed superiority of this program is not sufficiently supported by the contents or other evidence, what accounts for its acceptance?

To understand why the modern mathematics curriculum rather than some wiser version was promulgated it is necessary to note first the interests which modern mathematicians pursue. There is no question that up to the late nineteenth century the chief concern of the great mathematicians was to understand the workings of nature. We need not review here the relevant history because the assertion is not disputed. Mathematics was regarded as one of the sciences and indeed during the seventeenth, eighteenth, and most of the nineteenth centuries the distinction between mathematics and theoretical science was rarely noted. In fact, many of the men who have been ranked as the leading mathematicians of the past did far greater work in astronomy, mechanics, hydrodynamics, elasticity, and electricity and magnetism. Mathematics was simultaneously the queen and the handmaiden of the sciences.

Nor did these men hesitate to put to practical use the scientific knowledge that they and others had gathered. Newton studied the motion of the moon to help sailors determine their longitude at sea. Euler studied the design of ships and of sails, made maps, and wrote a masterful text on artillery. Descartes designed lenses to improve the telescope and microscope. Gauss not only made a survey of the electorate of Hannover, but worked on the improvement of the electric telegraph and the measurement of magnetism. These few examples could be multiplied a hundredfold. Almost all of these men not only saw the potential in the scientific knowledge they were helping to amass but were keenly concerned that the knowledge be utilized.

However, most mathematicians of the past hundred years have broken away from science. They know no science, and what is more, are no longer concerned with the utilization of mathematical knowledge. It is true that some, aware of the noble tradition that motivated mathematical research in the past and that warranted the honor accorded to men such as Newton and Gauss, still claim potential scientific value for their mathematical work. They speak of creating models for science. But in truth they are not concerned with this goal. In fact, since most modern mathematics professors know no science they can't be creating models. They are quite willing to shine by reflection of the light shed by great mathematicians of the past and even justify support of their present research by citing the accomplishments of their predecessors. Mathematics now is turned inward; it feeds on itself; and it is extremely unlikely, if one may judge by what happened in the past, that most of the modern mathematical research will ever contribute to the advancement of science. When confronted with this charge, the mathematicians dare not deny it—but then defend their creations on the ground that they are beautiful. Whether or not there is beauty in the creations need not be argued here. The important point is that this value is used to justify the work.

Another feature of the current mathematical activity is the narrow specialization. Mathematics has expanded enormously, as has science, and most mathematicians are almost obliged to concentrate on limited areas in order to keep abreast of other people's creations and produce new results of their own. Needless to say, the training of new mathematicians, which is conducted by professors who are themselves specialists in narrow fields, follows the same course. Doctoral candidates are forced to burrow into obscure corners in order to produce satisfactory theses. They are no longer broadly educated in mathematics, to say nothing of science.

Emphasis on mathematics proper and on specialization is especially strong in the United States. The primary reason is that in this country research is a relatively new phenomenon, and American professors anxious to shine and to train students who will shine, specialize in order to produce results quickly. The classical mathematical efforts, which have scientific goals, require extensive background because the subjects involved have been explored for several hundred years. Consequently, only a small percentage of the mathematicians, those often labeled applied mathematicians, continue to pursue the traditional goals. Most have turned to purely mathematical problems and to the formalization, axiomatization and generalization of what is already known. Such tasks are far easier.

The break between mathematics and science was deplored by the famous mathematical physicist John L. Synge as far back as 1944.

> Most mathematicians [today] work with ideas which, by common consent, belong definitely to mathematics. They form a closed guild. The initiate forswears the things of this world and generally keeps his oath. Only a few mathematicians roam abroad and seek mathematical sustenance in problems arising directly out of other fields of science. In 1744 or 1844 this second class included almost the whole body of mathematicians. In 1944 it is such a small fraction of the whole that it becomes necessary to remind the majority of the existence of the minority, and to explain its point of view.

> The minority does not wish to be labelled "physicist" or "engineer," for it is following a mathematical tradition extending through more than twenty centuries and including the names Euclid, Archimedes, Newton, Lagrange, Hamilton, Gauss, Poincaré. The minority does not wish to belittle in any way the work of the majority, but it does fear that a mathematics which feeds solely on itself will in time exhaust its interest. . . . Out of the study of nature there have originated (and in all probability will continue to originate) problems far more difficult than those constructed by mathematicians within the circle of their own ideas. . . .

> At present science is humming as it never hummed before. There are no obvious signs of decay. Only the most observant have noticed that the watchman has gone off duty. He has not gone to sleep. He is working as hard as ever, but now he is working solely for himself. . . .

Change and death in the world of ideas are as inevitable as change and death in human affairs. It is certainly not the part of a truth-loving mathematician to pretend that they are not occurring when they are. It is impossible to stimulate artificially the deep sources of intellectual motivation. Something catches the imagination or it does not, and if it does not, there is no fire. If mathematicians have really lost their old universal touch—if, in fact, they see the hand of God more truly in the refinement of precise logic than in the motion of the stars—then any attempt to lure them back to their old haunts would not only be useless—it would be denial of the right of the individual to intellectual freedom.

Some mathematicians, instead of pretending concern for the utility of their work, brazen forth a new declaration of independence. Professor Marshall H. Stone, then at the University of Chicago, in his article "The Revolution in Mathematics," decided to take the bull by the horns.

While several important changes have taken place since 1900 in our conception of mathematics or in our view concerning it, the one which truly involves a revolution in ideas is the discovery that mathematics is entirely independent of the physical world. To put this just a little more precisely, mathematics is now seen to have no necessary connections with the physical world beyond the vague and mystifying one implicit in the statement that thinking takes place in the brain. . . . When we stop to compare the mathematics of today with mathematics as it was at the close of the nineteenth century we may well be amazed to note how rapidly our mathematical knowledge has grown in quantity and in complexity, but we should also not fail to observe how closely this development has been involved with emphasis on abstraction and an increasing concern with the perception and analysis of broad mathematical patterns. . . . We realize too that the trend toward abstraction must inevitably continue, reinforced by the successes which are already to be credited to it. In following this trend and directing their attention more and more to the discernment and study of abstract patterns, mathematicians have become increasingly aware of the fundamental antithesis between the structural aspect of mathematics and the strictly manipulative aspect which so often appears to have paramount importance for the applications and so often is the principal preoccupation of the mathematics teacher. . . .

A modern mathematician would prefer the positive characterization of his subject as the study of general abstract systems, each one of which is built of specified abstract elements and structured by the presence of arbitrary but unambiguously specified relations among them. He would mean by the study of such mathematical systems not only the examination of intrinsic properties of individual systems but also the comparison of the structures of different systems. He would maintain that neither these systems nor the means provided by logic for studying their structural properties have any direct, immediate, or necessary connections with the physical world.

. . . Indeed, it is clear that mathematics may be likened to a game—or rather an infinite variety of games—in which the pieces and moves are intrinsically meaningless and the absorbing interest lies in perceiving and utilizing the patterns of play allowed under the rules. When mathematics is viewed in this light, the questions just noted pose the problem of determining whether or not it is possible to reduce the play of one or another of these games to a prescribed automatic procedure, leaving no room for the exercise of judgment and inspiration.

The views expressed by Stone and others have not gone unopposed. Richard Courant, formerly head of the mathematics department at the pre-Hitler world's center for mathematics, the University of Göttingen, and then head of the Courant Institute of Mathematical Sciences of New York University, replied to Stone (see the reference to Carrier in the bibliography),

Many of you have seen in the last issue of the *American Mathematical Monthly* [October, 1961] an article which has particular relevance for our present panel discussion. The article (by Professor Marshall Stone) discusses "The Revolution in Mathematics"; it asserts that we live in an era of great mathematical successes, which outdistance everything achieved from antiquity until now. The triumph of "modern mathematics" is credited to one fundamental principle, abstraction and conscious detachment of mathematics from physical and other substance. Thus, the mathematical mind, freed from ballast, may soar to heights from which reality on the ground can be perfectly observed and mastered.

I do not want to distort or belittle the statements and the pedagogical conclusions of the distinguished author, but as a sweeping claim, as an attempt to lay down a line for research and before all for education, the article seems a danger signal, and certainly in need of supplementation. The danger of enthusiastic abstractionism is compounded by the fact that this fashion does not at all advocate nonsense, but merely promotes a half truth. One-sided half truths must not be allowed to sweep aside the vital aspects of the balanced whole truth.

Certainly mathematical thought operates by abstraction; mathematical ideas are in need of abstract progressive refinement, axiomatization, crystallization. It is true indeed that important simplification becomes possible when a higher plateau of structural insight is reached. Certainly it is true—and has been clearly emphasized for a long time—that basic difficulties in mathematics disappear if one gives up the metaphysical prejudice of mathematical concepts as descriptions of a somehow substantive reality.

Yet, the life blood of our science rises through its roots; these roots reach down in endless ramification deep into what might be called reality, whether this "reality" is mechanics, physics, biological form, economic behavior, geodesy, or for that matter, other mathematical substance already in the realm of the familiar. Abstraction and gen-

eralization are not more vital for mathematics than indi
viduality of phenomena and, before all, not more than in
ductive intuition. Only the interplay between these forces
and their synthesis can keep mathematics alive and prevent its drying out into a dead skeleton. We must fight
against attempts to push the development one-sidedly to-
wards the one pole of the life-spending antinomy.

We must not accept the old blasphemous nonsense that
the ultimate justification of mathematical science is the
"glory of the human mind." Mathematics must not be allowed to split and to diverge toward a "pure" and an "applied" variety. It must remain, and be strengthened as, a
unified vital strand in the broad stream of science and must
be prevented from becoming a little side brook that might
disappear in the sand. . . .

Perhaps the most serious threat of one-sidedness is to
education. Inspired teaching by broadly informed competent teachers is more than ever an overwhelming need for
our society. True, curricula are important; but the cry for
reform must not be allowed to cover the erosion of substance, the propaganda for uninspiring abstraction, the
isolation of mathematics, the abandonment of the ideals of
the Socratic method for the methods of catechetic dogma-
tism.

. . . At any rate, it would be without doubt a radical
and vitally needed remedy for many ills in our schools and
colleges if a close interconnection between mathematics,
mechanics, physics and other sciences would be recognized
as a mandatory principle which must be vigorously em-
braced by the coming generation of teachers. To help such
a reform is a solemn obligation of every scientist.

On another occasion, in his necrology on a well-known
mathematician, Courant expressed concern about the
neglect of science. "There exists the danger that the applied mathematics of the future must be developed by
physicists and engineers and that professional mathematicians of rank will have no contact with the new developments."

The trend toward abstraction, toward mathematics for mathematics' sake, led the world-renowned mathematician John von Neumann to issue a warning. In his essay "The Mathematician" he stated,

> As a mathematical discipline travels far from its empirical source, or still more, if it is a second and third generation only indirectly inspired by the ideas coming from "reality," it is beset with very grave dangers. It becomes more and more pure aestheticizing, more and more purely *"l'art pour l'art."* This need not be bad if the field is surrounded by correlated subjects which still have closer empirical connections or if the discipline is under the influence of men with an exceptionally well-developed taste. But there is grave danger that the subject will develop along the line of least resistance, that the stream, so far from its source, will separate into a multitude of insignificant branches, and that the discipline will become a disorganized mass of details and complexities. In other words, at a great distance from its empirical source, or after much "abstract" breeding, a mathematical subject is in danger of degeneration. At the inception the style is usually classical; when it shows signs of becoming baroque, then the danger sign is up.

Still another protest was voiced in 1962 by a prominent mathematician, Professor James J. Stoker of New York University:

> It is a strange thing that in this country, which prides itself so much on the practical use it makes of all the scientific knowledge which has been gathered over the centuries, mathematics has been pursued for the last fifty years or more in a pronouncedly abstract manner and that that side of our science in which the relation between mathematics and the physical world plays an important role has been very much neglected. . . .
>
> . . . I observe that the abstract point of view and the neglect, even the contempt, for that kind of mathematics

which concerns itself with the world of reality, still represents the prevailing tone in American mathematics. The
plain fact is that the leading practitioners of that branch
of mathematics in which the interplay with mechanics and
physics is a strong motivation are nearly all of the older
generation, and there seem to be a very few replacements
for them in sight. In my view this is not a healthy state—
neither for our science itself, nor for the welfare of this
country. Furthermore there are strong forces at work, I
observe, which have the tendency to perpetuate this situation by propagating the notion that the strongly abstract
approach to mathematics is the suitable way to introduce
it to children in the elementary schools. It would seem to
me that this attitude ignores human psychology and turns
reason upside down: it ignores the historical fact that the
mode of progress in mathematics has always consisted in
formulating the appropriate and truly valuable abstractions
on the basis of prolonged experience of a very concrete
character, and the accompanying highly plausible infer-
ence that that is also the way most people's minds work.

It should not be thought that I and other colleagues who
share my attitude feel ourselves in opposition to those
mathematicians who choose to pursue their work in as
abstract a fashion as they find suitable and rewarding. Our
point of view is that it is vital for the health of our science
that the contact with the physical world should be preserved and cultivated, not merely because of the obvious
practical achievements which inevitably result from such
work, but because the whole history of mathematics shows
that such preoccupations have a stabilizing, vitalizing, and
fruitful effect on our science.

The danger to mathematics of the break from science
has been stressed by many other men. The leading American mathematician George D. Birkhoff, professor of
mathematics at Harvard University, said as far back as
1943: "It will probably be the new mathematical dis-
coveries which are suggested through physics that will
always be the most important, for, from the beginning

Nature has led the way and established the pattern which mathematics, the language of Nature, must follow."

To the argument that mathematics is now potentially more powerful for science because it is free to follow its own course, the "applied mathematicians" counter with the evidence of history. All the applications of mathematics to science came from mathematical ideas which were inspired by science. No mathematician ever cooked up ideas useful to science by sitting in an ivory tower. It is true that ideas inspired by science later found unexpected application, but the ideas were sound to start with because they derived from genuine physical problems. In the article already mentioned Synge remarks on this point too.

> Nature will throw out mighty problems but they will never reach the mathematician. He may sit in his ivory tower waiting for the enemy with an arsenal of guns, but the enemy will never come to him. Nature does not offer her problems ready formulated. They must be dug up with pick and shovel, and he who will not soil his hands will never see them.

What does the nature of current mathematical research have to do with curriculum reform? The relevance lies in the fact that the leaders in this reform have been college professors. These men were educated in a mathematical world that has departed radically from the concept of mathematics that animated the great mathematicians of the past. About eighty-five per cent of the Ph.D.'s in mathematics are not only narrow specialists but are concentrated in corners of mathematical logic, algebra and topology, fields which, on the whole, are remote from science. These men do not know even freshman physics nor have they any desire to know it. Because they have no idea of the role that mathematics has played in history they are ignorant as mathematicians and certainly as educated human beings. Most present-day

professors pursue abstractions, generalizations, structure
rigor, and axiomatics. Since this is what most mathematicians do, it is not surprising that this is what they
think mathematics education should train young people
to do. These college professors, competent or incompe-
tent, when called upon to help in the preparation of curricula, can suggest as subject matter only the narrow,
specialized abstract topics that they are familiar with or,
making some concession to the elementary level of instruction, they suggest somewhat watered-down or abbreviated versions of the more sophisticated treatments
of traditional mathematics. This fact accounts in large
part for the content of the modern mathematics curricula.

Professor Stone, whose characterization of modern
mathematical research was described previously, did not
hesitate to say in the same article that the curriculum
must be refashioned to teach that kind of mathematics.
"It is entirely obvious that these new insights and advances, which in sum constitute a genuine revolution in
mathematics, pose difficult practical problems for the educator. Merely to incorporate into the mathematical curriculum the essential elements of our new mathematical
knowledge is a formidable enough task, but the necessity
for presenting mathematics as the abstract subject it has
become and reconciling its antithetical aspects greatly
increases the difficulties involved in bringing mathematical instruction up to the level demanded by our
times. . . ."

The consequences of having university professors lead
curriculum reform are even more harmful. It is generally
conceded that college professors are chosen largely for
their knowledge of subject matter and research strength
and not for their pedagogical skill. Trained to do research, they are ill-prepared for teaching even on the college level. Mathematicians are not pedagogues. In fact
the two classes are almost disjoined sets. It did not occur

to these men that the goals of elementary, high school and even undergraduate education and the interests and capacities of students at these levels have little to do with mathematical research. Having become wise through the acquisition of a Ph.D. and possibly a prestigious position at a major university they believe themselves to be experts in areas in which they are in fact totally ignorant. Despite pedagogical failings when teaching on their own level and despite the fact that most of the professors who participated in curriculum reform had not been inside an elementary or high school since their own student days, the mathematics professors did not hesitate to take on a task that calls for considerable pedagogical acumen. One can say that they were presumptuous. They acted as though pedagogy was only a detail, whereas if they had really learned anything at all from their studies, they would have known that almost any problem involving human beings is enormously complex. The problems of pedagogy are indeed more difficult than the problems of mathematics, but the professors had supreme confidence in themselves. The trouble with most men of learning, as one wit put it, is that their learning goes to their heads.

Dr. Alvin M. Weinberg, Director of the Oak Ridge National Laboratory, in his article "But Is the Teacher Also a Citizen?" criticized the narrow professional point of view of mathematicians and scientists. Speaking of both groups he said,

> Thus, our science tends to become more fragmented and more narrowly puristic because its practitioners, harried as they are by the social pressures of the university community, have little time or inclination to view what they do from a universe other than their own. They impose upon the elementary curricula their narrowly disciplinary point of view, which places greater value on the frontiers of a field than on its tradition, and they try to put across what seems important to them, not what is im-

portant when viewed in a larger perspective. The practitioners have no taste for application or even for interdisciplinarity since this takes them away from their own universe; and they naturally and honestly try to impose their style and their standards of value. . . .

At another place in the article Weinberg noted the trend in both the new science and mathematics curricula.

But insofar as the new curricula have been captured by university scientists and mathematicians of narrowly puristic outlook, insofar as the curricula reflect deplorable fragmentation and abstraction, especially of mathematics, insofar as the curricula deny science as codification in favor of science as research, I consider them to be dangerous. . . .

. . . The professional purists, representing the spirit of the fragmented, research-oriented university, got hold of the curriculum reform and, by their diligence and aggressiveness, created puristic monsters. But education at the elementary level of a field is too important to be left entirely to the professionals in that field, especially if the professionals are themselves too narrowly specialized in outlook.

It is perhaps unnecessary to add that the professional mathematicians are so intent on making their careers through mathematical research that they take little or no time to acquire any knowledge of the history of their subject or of its human and cultural significance. Some even boast of their ignorance of science. A few may be informed in the broader values of mathematics but do not think it is necessary to teach it. Hence mathematicians are not really prepared to put their subject in an interesting light and thereby attract to the subject students who might very well take to it if the classroom material were appealing. Even if anxious to attract students, the limited professors are unable to do so and will not make the necessary effort to acquire the proper background.

Professor Feynman, whose article "New Textbooks for the New Mathematics" has already been cited, also scathingly criticized the new mathematics because the texts were written by pure mathematicians who are not interested in the connections of mathematics with the real world nor in the mathematics used in science and engineering because it is on the whole not new but old.

None of the above criticisms is intended to challenge the good intentions of the college professors but good intentions often succeed in doing no more than paving certain roads. Nevertheless their mistakes were so gross that one cannot but ask, "How could they have gone so far wrong?" In part, we have already accounted for their errors, but there is more to the story. As professionals with extensive training in mathematics they had acquired some understanding of the subject. Forgetting that they themselves had required years to achieve this understanding they believed that they could impart it at once to young minds. Moreover, their interest was to develop future mathematicians, but because they overlooked the pedagogy they failed even in that task. They concentrated on the superficial aspect of mathematics, namely, the deductive pattern of well-established structures, instead of emphasizing how to think mathematically, how to create and how to formulate and solve problems. Moreover, professional mathematicians are already motivated to pursue mathematics. Hence they failed to take into account that other people do not see the point of studying mathematics.

The professional mathematicians are the most serious threat to the life of mathematics, at least so far as the teaching of the subject is concerned. They resent students who do not take to the subject at once and are impatient with students who want to be convinced that the subject is worthy of interest. Yet mathematics proper can be a deadly subject especially because the courses are arranged in the order determined by the proper logical sequence and this means much drudgery to progress in the subject.

If mathematics professors were required to spend eight years in elementary school, three or four years in high school and then another year in college on ceramics they would object strenuously. But they do not see that mathematics for mathematics' sake has even less appeal to young people than ceramics has for themselves.

Mathematicians of this century are very much concerned with rigor. There are historical reasons for this preoccupation. However, we have already pointed out how damaging rigorous proof is to the student. Why does the mathematician insist on it? The answer is that he is favoring his professional interest, as, for example, in building on a minimal set of axioms. He is not willing to consider the pedagogy.

To prepare curricula at any level one must know the objectives of education at that level. For example, it is far more important in the lower grades to interest students in learning than it is to develop proficiency in any one subject. To know these objectives one must devote a great deal of thought to the whole problem of education. And this mathematicians do not and will not do.

Most mathematicians are not at all interested in the psychology of learning. This is a very difficult subject, more difficult than mathematics. How much can young people learn? Should they be induced by an offer of candy to learn something by rote? Do abstractions come easier to young people? Would negative numbers seem less artificial if taught early? A pedagogue will do his best to find out what psychologists have established and also learn from his own experience. Mathematicians will not take the trouble to find out what psychology can offer, nor will they take the trouble to develop their own skill in the art of teaching.

It is also easy to see why the texts are so poorly written. Professional mathematical writing has a style of its own. It is succinct, monotonous, symbolic and sparse. The chief concern is to be correct. On the other hand, good texts must have a lively style, arouse interest, tell students where they are going and why. Writing is an art and mathematicians do not cultivate it.

One of the basic reasons that mathematicians fail as pedagogues stems from the nature of the mathematical mind. The common belief is that the mathematician is the epitome of intelligence and hence should always be able to act wisely and to prescribe solutions to all problems. Most people believe this because they are scared by the mere appearance of symbols and conclude that if a man can master these symbols he must be intelligent. One might just as well conclude that every Frenchman must be intelligent because he masters French. But I shall venture to draw a distinction between a mathematician's intellectual capacities and his wisdom. The mathematician does have the ability to make sharp distinctions in the meanings of words, the capacity to learn and apply the laws of logic, and the capacity to retain and compare a number of facts. He has what I shall call a rational mind. He may also be creative in mathematics. Wisdom may indeed include these rational qualities but it also includes much more: judgment, the capacity to learn from experience, the perception of values, the understanding of human beings, and the capacity to use knowledge for the solution of human problems. These latter qualities are not possessed by mathematicians any more than by any other group of people selected at random.

It is rational to present mathematics logically, but it is not wise. Consider the teaching of calculus. One knows that the calculus is built on the theory of limits and so may conclude that the way to teach calculus is to start with the theory of limits. The wise man would also consider whether young people can learn the theory of limits from scratch and whether they will want to learn it without motivation and prior insight.

On a sheer probability basis, wisdom would be distributed among mathematicians as among lawyers, doctors, engineers and businessmen. But a little analysis seems to raise the question of whether mathematicians are likely to have their share. What attracts people to mathematics? Mathematics is a simple subject compared to economics, psychology or physics. It is a narrow subject. One does not have to have an extensive background to do pure mathematics. Moreover, mathematics *per se* does not deal with human beings and the complex problems that dealing with human beings pose. As Bertrand Russell said, "Remote from human passions, remote even from the pitiful facts of nature, the generations have created an ordered cosmos, where pure thought can dwell as in its natural home and where one, at least, of our nobler impulses can escape from the dreary exile of the actual world." Hence mathematics is likely to attract those who do not feel competent to deal with people, those who shy away from the problems of the world and even consciously recognize their inability to deal with such problems. Mathematics can be a refuge.

Mathematicians as a class are overrated in another essential respect. One tends to assume that professors are superior in character and that they will therefore espouse only those causes and movements that are helpful to society. Unfortunately, mathematics has its share of opportunists, bandwagon-jumpers, reactionaries, prestige-seekers, power-seekers, and money-grabbers. It is sad to read in the history of mathematics that even many great mathematicians stooped to presenting as their own results they took from others.

This appraisal of mathematicians is terribly negative, but it seems necessary to discredit the belief that mathematics professors are infallible and a truly superior group.

In view of their overriding concern with personal advancement through research, their unwillingness to devote time to the problems of pedagogy, and possibly limited development as wise human beings, it is not likely that college professors can lead curriculum work on the high school and elementary school levels.

How did college professors become involved in elementary and high school curriculum reform? There is no question that some help from this source was needed. Elementary and high school teachers do not have enough time to follow the advances in mathematics, to keep abreast of any new material that should be integrated into the curriculum, to profit from any investigations in the learning process, and to incorporate such material in any large-scale curriculum revision. They need to be apprised of these matters by professors, whose function it is to be informed in these areas. Hence the mathematics professors were called upon to participate in reform.

There is another group that might have served usefully in curriculum reform. This is the group of education professors. However, most of these men were not informed in advanced mathematics and concentrated on how to teach the traditional mathematics. The plight of these professors was expressed openly and honestly by Professor Max Beberman, who was a mathematics-education professor. In the Proceedings of a University of Illinois Committee on School Mathematics Conference held at the University of Illinois in 1964 and devoted to

"The Role of Applications in a Secondary School Mathematics Curriculum," he said,

> My point of view has always been that I, personally, have very little responsibility for the selection of the content to be tried out experimentally. My job is to find out what things can be taught and what things can't be taught. So, if someone makes a suggestion about a topic to be taught and it turns out that, when I give my best efforts to this, I still can't get it across to children, maybe it can't be taught—maybe. I don't know how much harm we do to students if we select good mathematical content in the first place and then exert our best efforts to get it across. I'm perfectly content, as a professor of Education, to devote all of my attention to finding the right kind of pedagogy to get mathematical ideas across to children, but many critics have pointed out that you can't trust mathematicians with the selection of content—that, somehow or other, mathematicians don't understand what high school students can learn. This objection seems silly to me because who knows what students can learn? That is something for experimentation. A more serious protest has been that mathematicians don't know just what is appropriate mathematics for students. They don't know what the really important things are in mathematics as far as general education is concerned. Now I don't know on whom I am supposed to call for this kind of insight. Should we call on people who are not expert in mathematics to tell us what mathematics is appropriate? This, I think, is a problem with which we are always going to be faced. I am still willing to ask mathematicians for help in this regard. I don't think that they have misled us too much in the past. But I think it is important to recognize that there are a lot of people now, more than there were three or four years ago, who question the assumption that mathematicians are the best people to advise on what mathematics to teach in the high school.

The appeal to professors of mathematics for information on what subjects should be pursued is not in itself a mistake and in fact, as we have pointed out, is necessary. Unfortunately, the ones who participated in curriculum work were not in the main the choice ones. Those few who still worked in the traditional areas of mathematics, the areas concerned with mathematics and its relationship to the sciences, were in the 1960s heavily engaged in research on the problems of the rapidly expanding sciences. Hence it was the professors from the more remote, abstract and pure side of mathematics who felt free to devote themselves to curriculum work. Moreover, apart from their own onesidedness, these men, like college professors generally, had had no experience or contact with elementary and high school curricula. In fact, they had disdained such interests in the past and so had no idea of what should be taught at these levels or how young people think. Many professors of questionable competence, noting that they would have to deal only with elementary mathematics and seeking some activity that might add to their prestige, gladly joined in.

One might think that the pedagogical weaknesses of the college professors would be offset by the high school teachers and the education professors. Surely the latter two groups should know what can be taught to elementary and high school students and what might motivate these young people. But the high school teachers and the professors of education were overawed by the mathematicians. One who can write a research paper is regarded in our society as a person of extraordinary ability. How could lowly members of the fraternity question such men of distinction and unquestionable knowledge? It seems fair to say on the basis of what transpired that in every curriculum group the college professors dominated the educators and the schoolteachers. The educators and teachers bowed down to the idols, not knowing that most had feet of clay.

This account of what transpired does not argue against

the involvement of intellectuals in the problems of the schools. The need for the involvement has been acknowledged. But it may show that school systems have to be strong enough to recognize when they are being helped and when they are being led astray. The value of the aid the schools will get will depend entirely on the competence of those being helped. They must judge whether the advice they get is sound. Put otherwise, college professors can be used as consultants but certainly should not lead and dominate the fashioning of curricula for the elementary and high schools.

We have accounted for the direction which the new mathematics curriculum took, but this does not explain its relatively widespread acceptance. In view of the manifold defects of this curriculum one would think that it would be rejected outright by the country at large. Several factors account for its adoption.

Probably the largest single factor is that the curriculum groups were organized and well financed. Hence these groups undertook active campaigns to put the new curricula across. Not only the leaders but members of the many groups began to speak for the new curricula at various meetings of teachers, principals and administrators. Since the traditional curriculum was not successful, these people were at least receptive to a new curriculum. When assured that mathematicians, education professors and high school teachers had collaborated and were agreed on the merits of the new curriculum, those addressed were impressed.

Beyond speeches, the curriculum groups issued literature. The document already mentioned, *The Revolution in School Mathematics,* and subtitled, A Challenge for Administrators and Teachers, was issued by the National Council of Teachers of Mathematics in 1961. Ostensibly the document was a report on conferences held around the United States to inform school administrators and mathematics supervisors of the nature of the new curricula. But it described the curricula as ones "that would enable them to provide leadership in establishing new and improved mathematics programs." It implied that administrators who failed to adopt the reforms were guilty of indifference or inactivity. But in 1961 this country had had very little experience with the new curricula. A booklet championing and advocating them at that time can be fairly accused of propaganda.

Unfortunately, the propaganda was effective. Most school administrators do not have the broad scientific background to evaluate the proposed innovations. They do not know whether these innovations are models of educational know-how combined with superlative subject matter or are the enthusiasms of subject-matter specialists without much relevance to the needs of students. The pressure does put the administrators on the spot. They can seem to show interest and progress by adopting one of the modern programs, or they can be thoroughly honest and admit that they are not competent to judge the merits of any one. What actually happened is that many principals and superintendents urged the modern curricula on their teachers just to show parents and school boards that they were alert and active.

The very adoption of the term modern mathematics is pure propaganda. "Traditional" connotes antiquity, inadequacy, sterility, and is a term of censure. "Modern" connotes the up-to-date, relevant, and vital. The terms modern and new were used for all they were worth. Speakers capitalized on the fact that the traditional curriculum offered little that was not known before 1700. Of course, as we have seen, the terms modern and new were hardly justified since in the main the new curricula offer a new approach to traditional mathematics.

A few speakers actually degraded themselves by resorting to thinly veiled threats. Some of them were on the Commission on Mathematics of the College Entrance Examination Board. This Board formulates the aptitude tests which high school students take for admission to college. The speakers hinted that these tests would contain questions on the modern mathematics topics. Since the teachers were anxious to have their students do well on these tests, they felt compelled to learn what the new curricula contain and compelled to teach these topics.

Another device, deliberately employed to put modern mathematics across, was described by Professor Paul Rosenbloom, a very competent mathematician, who participated in the writing of curricula. In the article, "Applied Mathematics: What is Needed in Research and Education" (see the reference to Carrier in the bibliography), Professor Rosenbloom described this device: "Now, in this business of trying to revise teaching of mathematics in the schools, we had many problems. Among them was the problem of social engineering, which was that you had to make a big change quickly under conditions where no one had the power to impose anything on anybody. You had to get a lot of people to decide voluntarily to do what they ought to do. And so it was not just a problem of figuring out what mathematics should be taught in grade 7 or whatever other grade. But there was also a problem of actually getting this into a large number of schools quickly without being able to force them. So, the procedure had to be one where a large number of people were assembled from different parts of the country and representing different points of view, both from mathematics and from education, from the schools and so on, to write books in a very permissive atmosphere. Essentially, the problem was to be able to say that this ninth-grade course, for example, represents the consensus of mathematicians and teachers. The Seattle school system has introduced SMSG into the Seattle school system because their mathematics supervisor was on the writing team. The National Council of Mathematics Teachers has had a voice in this. The chairman of their secondary school curriculum committee was on the eleventh-grade writing team, and so on. You have these problems, and to a certain extent, what came out depended upon the persuasiveness and the maturity of judgment of the people that got on the writing teams that represented various points of view." Such promotional schemes led one school consultant to say that if the pedagogical insights of the developers of some of the modern programs were equal to their promotional acumen, the millennium of mathematics education would be here.

Even though many teachers on the writing teams were disappointed and even chagrined by the compromises they were obliged to make, they nevertheless returned to their home districts proud to have been participants in fashioning curricula and naturally inclined to favor what they had helped to create. They soon became the ardent champions of modern mathematics and took the lead in promoting it.

Many college professors seeking an activity in which they believed they would be competent—surely, they argued, high school material is child's play for us knowledgeable college professors—took up the advocacy of the modern curricula and even initiated courses to train teachers in modern mathematics. Others climbed on the bandwagon because it gave them visibility and even prestige to participate in what had become a prominent activity. High school and elementary school teachers, too, anxious to show leadership in education, have taken up the promotion of what was thrust upon them. Unfortunately, because they are committed to full-time, exhausting work in their duties as teachers, they have not had the opportunity to learn more about what is significant in mathematics and so have not been able to examine critically the new versions as the road to education in mathematics.

Many teachers jumped in because they saw an opportunity to advance their writing efforts by catering to the new mathematics. Often they have done no more than dress up old texts with some sprinkling of the new mathematics and label these books modern mathematics texts. Naturally such teachers will at least outwardly profess advocacy of modern mathematics.

One could rationalize such texts as compromises. The argument would be that students are not ready for a radical shift to modern mathematics, especially if they have already had a few years of traditional mathematics. But these compromise texts are really not transitional. They have not worked out a reasonable unification of traditional and modern topics. They are clearly commercial jobs that give the impression of being modern mathematics texts but actually are patches of traditional and modern topics that are entirely unintegrated.

The publishers, seeking to gain the edge on the market, put out series of new mathematics texts and, to ensure their adoption, not only joined in the propaganda for the new mathematics through cleverly worded advertisements but sent speakers to teachers' meetings to speak for the new mathematics. The combination of curriculum leaders, teachers who became partisans, and publishers now form a tightly knit web of vested interests preying on the mathematical unsophistication of press, public and even foundations who support this movement.

Beyond the factors we have already described there are other reasons that modern mathematics is favored. There is no doubt that some teachers actually believe that the axiomatic deductive approach is the essence of mathematics. Whether they acquired this limited view through the instruction they themselves received or have been induced to adopt it because many texts favor it, they are at least sincere if not effective pedagogues. One also has the sneaking suspicion that a few teachers enjoy presenting the familiar number system in the recondite axiomatic form because they understand the simple mathematics involved and yet can appear to be teaching profound subject matter.

Many young teachers believe that, now that we have the correct polished version of mathematics, it is sufficient to give the axiomatic or rigorous approach and that students will absorb it. These very same teachers would have been swamped by such a presentation, but having learned the correct version they can no longer recall and appreciate the difficulties they encountered in learning the rigorous versions.

Some teachers, knowing the rigorous proofs, feel uneasy about presenting merely a convincing argument which they, at least, know is incomplete. But it is not the teacher who is to be satisfied; it is the student. Good pedagogy demands such compromises.

The natural desire of the teacher is to proffer the completed polished deductive mathematics. This is certainly the more elegant version. But the value to the student is inversely proportional to the elegance and smoothness of the organization, because the final version is a much reworked and unnatural account.

Other teachers want to give students the whole truth at once so that they should not have to unlearn what they once learned. But one cannot teach even English or history by starting at the top. The A that a high school student might earn for an English composition would most likely be rated C at the college level. Further, teaching $2 + 2 = 5$ and then having to correct it is one thing but teaching subtraction as "taking away" and then introducing the notion that -2 is the additive inverse to 2 is another. For a youngster the latter is verbiage.

Another major reason for the popularity of the axiomatic deductive approach is that it is easier to present. The entire body of material is laid out in a clear, clean-cut sequence and all the teacher has to do is repeat it. He has but to offer a canned body of material. I have heard teachers complain that many students, particularly engineers, wish to be told only how to perform the processes they are asked to learn and then want to hand back the processes. But the teachers who present the logical formulation because it avoids such difficulties as teaching discovery, leading students to participate in a constructive process, explaining the reasons for proceeding one way rather than another, and finding convincing arguments, are more reprehensible than the students who wish to avoid thinking and prefer just to repeat mechanically learned processes. Postulating properties has the advantage, as Bertrand Russell put it, of theft over honest toil. Pedagogically it is worse because the theft produces no gain in understanding.

We have already noted that many teachers, especially at the college level, prefer to present rigorous axiomatic approaches because they favor their own professional interest at the expense of the student. Even if such systems could be made understandable to young people the time required to teach them should be spent on more significant material. In this matter as well as in presenting sophisticated rigorous proofs the teachers are serving themselves not only in the form in which they present mathematics but also in the premature teaching of subjects such as abstract algebraic concepts, linear vector spaces, finite geometries, set theory, symbolic logic and matrices, because these subjects are advanced and satisfy the teacher's ego. Is it any wonder that students become alienated and question the relevance of what they are being taught?

For whatever reason teachers insist on presenting to young people modern rigorous proof, they are deceiving themselves. As we have already noted (Chapter 5), there is no ultimate rigorous proof. This fact derives from the very way in which mathematics develops. The superb research mathematician and pedagogue Felix Klein has described it. "In fact, mathematics has grown like a tree, which does not start at its tiniest rootlets and grow merely upward, but rather sends its roots deeper and deeper at the same time and rate that its branches and leaves are spreading upward. . . . *We see, then, that as regards the fundamental investigations in mathematics, there is no final ending, and therefore on the other hand, no first beginning, which could offer an absolute basis for instruction.*" Poincaré expressed a similar view. There are no solved problems; there are only problems that are more or less solved. Mathematics is as correct as human beings are and humans are fallible.

At no time in the history of mathematics have we been less certain of what rigor is. Hence no proof is really complete and the teacher must compromise in any case. It would be interesting to know how many teachers are aware that set theory, which they now regard as the indispensable beginning to any rigorous approach to mathematics, has been the source of our deepest and thus far insuperable logical difficulties. Those who are not aware of the foundational problems might at least note the words of one of the foremost mathematicians of our time, Hermann Weyl. "The question of the ultimate foundations and the ultimate meaning of mathematics remains open; we do not know in what direction it will find its final solution nor even whether a final objective answer can be expected at all. 'Mathematizing' may well be a creative activity of man, like language or music, of primary originality, whose historical decisions defy complete objective rationalization."

Though the rigorous axiomatic deductive approach is favored, there are indications that some professors who present rigorous material are really uncertain as to the wisdom of doing so. A number of calculus books begin with rigorous definitions and theorems, for example, those concerning limits and continuity, and then never refer to this material. Thereafter they use the cookbook presentation. The most charitable view of such books is that the authors wish to ease their own consciences or to give the students some idea of what rigor means. Perhaps a more accurate view is that these books offer only a pretense of rigor in order to appeal to both markets, the one that demands rigor and the one that is satisfied to teach mechanical procedures.

Other texts adopt another "compromise." In the body of the text the presentation is mechanical with perhaps an occasional condescension to an intuitive explanation. The "real explanation" is given in rigorous proofs, but these are put in appendices and presented so compactly that they are certain to be totally un-understandable to the student. However, the authors have salved their consciences. Such books are no different from the old mechanical presentations. They do contribute to understanding in one respect, namely, they show that competent mathematicians are inept in pedagogy.

For the various reasons we have been citing, modern mathematics has now become quite fashionable. But fashion, as Oscar Wilde put it, is that by which the fantastic becomes for the moment universal.

11

The Proper Direction for Reform

"Logic can be patient for it is eternal."

Oliver Heaviside

We have shown that the traditional curriculum is defective in a number of respects and that the new mathematics curriculum certainly does not remedy the defects of the traditional curriculum. In addition, it introduces new defects. What direction, then, should effective reform take? Put roughly for the moment, the direction should be diametrically opposite to that taken by the new mathematics.

Before we can consider the approach and contents of a suitable elementary and high school curriculum, we must consider the objectives or goals of these stages of education. On the elementary school level there can be no consideration of preparation for college. Only a small percentage of these students will go to college. Even on the high school level, from which about fifty per cent of present day graduates enter college, the students are still

173

ignorant of the nature and importance of the various subjects they are asked to take. For many subjects, including mathematics (beyond arithmetic), the high school offering is an introduction. Moreover, very few of the college-bound students will specialize in mathematics. Even those who think they will become mathematicians should be advised not to specialize until they know much more about what the various subjects have to offer. Hence the education for all these students should be broad rather than deep. It should be a truly liberal arts education wherein students not only get to know what a subject is about but also what role it plays in our culture and our society. Put negatively, there should be no attempt to train professionals in mathematics and little concern for what future study in mathematics may require. In view of these facts, what values, beyond the arithmetic of daily needs, can mathematics offer?

Mathematics is the key to our understanding of the physical world; it has given us power over nature; and it has given man the conviction that he can continue to fathom the secrets of nature. Mathematics has enabled painters to paint realistically, and has furnished not only an understanding of musical sounds but an analysis of such sounds that is indispensable in the design of the telephone, the phonograph, radio, and other sound recording and reproducing instruments. Mathematics is becoming increasingly valuable in biological and medical research. The question "What is truth?" cannot be discussed without involving the role that mathematics has played in convincing man that he can or cannot obtain truths. Much of our literature is permeated with themes treating mathematical accomplishments. Indeed, it is often impossible to understand many writers and poets unless one knows what influences of mathematics they are reacting to. Lastly, mathematics is indispensable in our technology.

Should such uses and values of mathematics be taught in mathematics courses? Certainly! Knowledge is a whole and mathematics is part of that whole. The subject did not develop apart from other activities and interests. To teach mathematics as a separate discipline is a perversion, a corruption and a distortion of true knowledge. If we are compelled for practical reasons to separate learning into mathematics, science, history and other subjects, let us at least recognize that this separation is artificial and false. Each subject is an approach to knowledge and any mixing or overlap where convenient and pedagogically useful, is desirable and to be welcomed.

What we should be fashioning and teaching, then, beyond mathematics proper, are the relationships of mathematics to other human interests—in other words, a broad cultural mathematics curriculum which achieves an intimate communion with the main currents of thought and our cultural heritage. Some of these relationships can serve as motivation; others would be applications; and still others would supply interesting reading and discussion material that would vary and enliven the content of our mathematics courses.

Can such material be introduced at the elementary and high school levels? Of course. In fact, if even the elementary levels of our subject did not have intimate relationships with the major and vital branches of our culture, the subject would not warrant an important place in the curriculum.

The need to relate mathematics to our culture has been stressed by Alfred North Whitehead, the deepest philosopher of our age and a man capable of the most exacting abstract thought. In his essay "The Aims of Education," written in 1912, Whitehead says,

> In scientific training, the first thing to do with an idea is to prove it. But allow me for one moment to extend the meaning of "prove"; I mean—to prove its worth. . . .

The solution which I am urging, is to eradicate the fatal disconnection of subjects which kills the vitality of our modern curriculum. There is only one subject-matter for education, and that is Life in all its manifestations. Instead of this single unity, we offer children Algebra, from which nothing follows; Geometry, from which nothing follows. . . .

Let us now return to quadratic equations. . . . Why should children be taught their solution? . . .

Quadratic equations are part of algebra and algebra is the intellectual instrument for rendering clear the quantitative aspects of the world.

In his essay of 1912, "Mathematics and Liberal Education" (published in *Essays in Science and Philosophy*), Whitehead goes further.

Elementary mathematics . . . must be purged of every element which can only be justified by reference to a more prolonged course of study. There can be nothing more destructive of true education than to spend long hours in the acquirement of ideas and methods which lead nowhere . . . there is a widely-spread sense of boredom with the very idea of learning. I attribute this to the fact that they [the students] have been taught too many things merely in the air, things which have no coherence with any train of thought such as would naturally occur to anyone, however intellectual, who has his being in this modern world. The whole apparatus of learning appears to them as nonsense. . . .

Now the effect which we want to produce on our pupils is to generate a capacity to apply ideas to the concrete universe. . . . The study of algebra should commence with a systematic study of the practical application of mathematical ideas of quantity to some important subject.

[In geometry, likewise, the curriculum] should be rigidly purged of all propositions which might appear to the student to be merely curiosities without important bearings. . . .

What, in a few words, is the final outcome of our

thoughts? It is that the elements of mathematics should be treated as the study of a set of fundamental ideas, the importance of which the student can immediately appreciate; that every proposition and method which cannot pass this test, however important for a more advanced study, should be ruthlessly cut out. . . . Again this rough summary can be further abbreviated into one essential principle, namely, simplify the details and emphasize the important principles and applications.

In 1912, Whitehead was addressing himself to the traditional curriculum, but the criticisms and positive recommendations apply with all the more force today.

Mathematics is not an isolated, self-sufficient body of knowledge. It exists primarily to help man understand and master the physical, the economic and the social worlds. It serves ends and purposes. We must constantly show what it accomplishes in domains outside of mathematics. We can hope and try to inculcate interest in mathematics proper and the enjoyment of mathematics but these must be by-products of the larger goal of showing what mathematics accomplishes.

Some men have gone so far as to recommend combining mathematics and science. Professor E. H. Moore, a noted mathematician, formerly of the University of Chicago, addressed himself in his paper "On the Foundations of Mathematics" to the problem of teaching mathematics and recommended combining mathematics and science on the high school level. He urged that the artificial separation of pure and applied mathematics be ended and that we build a continuous correlated program in secondary mathematics and science. In this way we might succeed in arousing in the learner "a feeling that mathematics is indeed a fundamental reality of the domain of thought, and not merely a matter of symbols and arbitrary rules and conventions."

Whether or not mathematics should be combined with

science, by presenting mathematics as a part of man's efforts to understand and master his world we would be giving students the historically and currently valid reason for the great importance of the field. This is also the primary reason for the presence of mathematics in the curriculum. We are therefore obliged to present this value of mathematics. Anything less is cheating the student out of the fruit of his learning.

By catering to the general student we do not in fact ignore the future professional. A small percentage of the students will be physicists, chemists, engineers, social scientists, technicians, statisticians, actuaries, and so on. It is desirable that these students get, as early as possible, some knowledge of how mathematics can help them in their future work. In fact, if they are already inclined toward one of these careers and if we show them how mathematics is useful in it, they will take an interest in the subject on behalf of their careers. Even if students do not already incline toward a particular career, we are obliged to open up the world to them and to make clear the nature of the various professions. One important way of doing this is to show how mathematics is involved in these fields.

To present mathematics as a liberal arts subject requires a radical shift in point of view. The traditional and modern approaches treat mathematics as a continuing cumulative logical development. Algebra precedes geometry because some algebra is used in geometry. Trigonometry follows geometry because a modicum of geometry is used in the former subject. The new approach would present what is interesting, enlightening, and culturally significant—restricted only by a slight need to include earlier concepts and techniques that will be used later. In other words, we should be objective-oriented rather than subject-oriented.

We have stressed so far that mathematics must be presented as an integrated part of a liberal education. Equally vital is another principle that should guide the presentation of mathematics proper, a principle disregarded by the traditional curriculum and even more so by the modern mathematics curriculum. Every mathematical topic must be motivated. Mathematics proper does not appeal to most students and they constantly ask the question, "Why do I have to learn this material?" This question is thoroughly justified.

Mathematics proper is a very limited subject. Hermann Weyl, one of the great mathematicians of our time and a man who worked in many branches of mathematics and mathematical physics, said in 1951, "One may say that mathematics talks about things which are of no concern at all to man. . . . It seems an irony of creation that man's mind knows how to handle things the better the farther removed they are from the center of his existence. Thus we are cleverest where knowledge matters least: in mathematics, especially in number theory."

Mathematics deals with abstractions and this in itself is one of its severest limitations. A discourse on the nature of man can hardly be as rich, as satisfying and as life-fulfilling as living with people, even though one may learn a great deal about people from an abstract discussion of man. To talk or to read about children is not to bring up a child. Beyond the fact that it is abstract the subject matter of these abstractions is remote. All of mathematics treats numbers and geometrical figures and generalizations arising from these basic concepts. But numbers and geometrical figures are insignificant properties of real objects. A rectangle may indeed be the shape of a piece of land or the frame of a painting but the shape is incidental to the real value of the land or the painting.

Mathematics proper does not and perhaps should not appeal to ninety-eight per cent of the students. It is an

esoteric study, entirely intellectual in its appeal and lack-
ing the emotional appeal of, say, music and painting.
The creative mathematician may derive some emotional
values such as satisfaction of the ego, pride in achievement, and glory—values none too noble in any case—
but the student cannot derive even these values from the
study of the subject, or if he does, the strength of these
emotions is slight. Intellectual challenge may arouse some
people, but one could hardly refute those who would
maintain that the challenges of building a more humane
society and securing honest leaders are more important.

Hence, motivation for the nonmathematician cannot
be mathematical. We have already noted that it is pointless to motivate complex numbers for the general student
by asking for solutions of $x^2 + 1 = 0$. Since nonmathe-
maticians don't care to solve $x - 2 = 0$ why should they
care to solve the former equation? Calculus texts "motivate" many of the concepts and theorems by applying
them to the calculation of areas, volumes and arc-lengths.
But these are also mathematical topics and the fact that
the calculus enables us to calculate them does not make
the subject more engrossing to the nonmathematician.
The pointlessness to the students of the many theorems of
Euclidean geometry has soured them on geometry; hence
more geometry, even if via the calculus, will arouse dis-
taste for the calculus rather than interest. The argument
that the calculus gives us a power to do what the more
elementary subjects cannot do certainly does not impress
students who don't wish to calculate areas, volumes and
arc-lengths in the first place.

The natural motivation is the study of real, largely
physical, problems. Practically all the major branches of
mathematics arose in response to such problems and certainly on the elementary level this motivation is genuine.
It may perhaps seem strange that the great significance
of mathematics lies outside of mathematics but this fact

must be reckoned with. For most people, including the great mathematicians, the richness and values that do attach to mathematics derive from its use in studying the real world. Mathematics is a means to an end. One uses the concepts and reasoning to achieve results about real things.

One would think that teachers on all levels should have long since recognized the deplorable lack of motivation, and instead of turning to new approaches to old subject matter and to new subject matter would have tackled the problem of motivation. There are, of course, some acceptable excuses for this failure. The average schoolteacher is obliged to follow a curriculum that is laid out for him. He may in his own class do something to inspire his students, but he is generally limited with respect to the time to do this. Moreover, revision so as to provide motivation would require a new approach. If it should lead to a new order of topics, he would not be allowed to depart from the syllabus and the material would be prohibited.

But participants in educational reform cannot be excused for their failure to supply motivation. The professors of mathematics education have been men of limited vision. Though their task is to teach how to present mathematics, they do not themselves know why mathematics is important and where it makes contact with real problems that might be used to motivate the student. Up to the time that the modern curriculum was fashioned, the college professors took little interest in the elementary and high school courses. They did take the lead in fashioning the new curriculum and we have already noted their shortcomings. On all these accounts the problem of motivation has not been met.

The motivation must be presented along with the topic to be taught. It will not do to assure students that they will some day appreciate the value of the mathematics they are asked to learn. If a subject has any value, then as Whitehead points out, the student must be able to appreciate its importance immediately; or as he put it in his *Aims of Education,* "Whatever interest attaches to your subject matter must be evoked here and now." If mathematics is not revivified by the air of reality we cannot hope that it will survive as an important element in liberal education.

Would real problems meet the interests of young people? They live in the real world and, like all human beings, either have some curiosity about real phenomena or can be far more readily aroused to take an interest in them than in abstract mathematics. Hence there is an excellent prospect that the genuine motivation will also be the one that interests students, and, indeed, some limited experience has shown this to be the case. But if it is not the complete answer, then further work must be done to secure the effective motivation. If puzzles, games, or other devices serve at particular age levels, these too can be used—though they cannot be the major source of motivation, else students will get the wrong impression of the value of mathematics. Certainly far more work can be done to cull motivating problems which would be genuine and meaningful to the student.

Motivation does not always call for preceding the treatment of a mathematical topic by a problem drawn from the sciences or real life. It is sometimes more convenient to introduce a mathematical topic, present the mathematics and then immediately apply it to a non-mathematical situation. For example, one of the topics of elementary geometry is parallel lines. One may present this and then show how a simple theorem enables us to calculate the circumference of the earth. The parabola as a curve may be taught as just a locus problem. But then the uses of the parabola in focusing and directing light and radio waves should surely be presented. Pic-

ures of automobile headlights, radio antennas, search-
lights and even the common flashlight show these uses
in real situations. In algebra we study linear and quadratic functions. These may be applied readily to calcu-
lating how high a ball or projectile directed straight up
will go and whether the projectile can reach a desired
height. In this age of space exploration shooting up rock-
ets and spaceships can be an absorbing topic even if only
the simplest problems can be considered at the high
school level.

By neglecting motivation and application the pedagogues have caused mathematics education to suffer.
These men have presented the stem but not the flower
and so have failed to present the true worth of what they
are teaching. They call upon students to fight battles but
do not tell them why they are engaged in them. Even
the United States Army knows better. Some of the
poorest teaching of mathematics is traceable to teachers
treating the subject as though it had no connection with
anything beyond its technical confines. What is especially
grievous, then, about the teaching of mathematics, traditional or new, is not that the teachers do not know what
they are teaching but that they do not know and so cannot show pupils why mathematics is vital.

There are many professors and teachers who feel that
motivation and application are a departure from the
legitimate content of mathematics courses. But knowing
what mathematics does is part of knowing mathematics.
Moreover, without motivation students do not take to
the mathematics proper and consequently little is accomplished by teaching just the mathematics. Plutarch said,
"The mind is not a vessel to be filled but a fire to be
kindled." Motivation kindles the fire.

The use of real and especially physical problems serves
not only to motivate mathematics but to give meaning to
it. Negative numbers are not just inverses under addition

to positive integers but are the number of degrees below 0° on a thermometer. The ellipse is not just a peculiar locus but the path of a planet or a comet. Functions are not sets of ordered pairs but relationships between real variables such as the height and time of flight of a ball thrown up into the air, the distance of a planet from the sun at various times of the year, and the population of a country over some period of years. Functions are laws of the universe and of society. Mathematical concepts arose from such physical situations or phenomena and their meanings were physical for those who created mathematics in the first place. To rob the concepts of their meaning is to keep the rind and to throw away the fruit.

Even one of the major curriculum groups, The Secondary School Curriculum Committee of the National Council of Teachers of Mathematics, stressed the value of applications in giving meaning to mathematics: "Applications of mathematics are important as media through which pupils might gain deeper appreciation of the tool value of mathematics and as aids in the clarification and illustration of mathematical content." Similarly, several of the individuals active in the modern mathematics curriculum work have spoken and written in favor of applications. One of these men went so far as to say that if in the study of mathematics there were no physical applications it would be necessary to invent some. But the texts are devoid of such applications.

There is another value to be derived from developing mathematics from real situations. One of the greatest difficulties that students encounter in mathematics is solving verbal problems. They do not know how to translate the verbal information into mathematical form. Under the usual presentations in the traditional and modern mathematics curricula this difficulty is to be expected. Mathematics is presented in and for itself, divorced from physical meaning, and then the students are called upon o relate this isolated, meaningless mathematics to real ituations. Clearly they have no foundation on which to ink about such situations. On the other hand, if the nathematics is drawn from real problems, the difficulty of ranslation is automatically disposed of.

So far as the actual manner of presenting mathematics s concerned, there is another principle that should be ollowed. Mathematics should be developed not deducively but constructively. Alternatively one says today hat one must teach discovery. The lip service paid to this rinciple would fill many volumes, but the practice could e encompassed in the empty set. The constructive approach means that the students should do the building of he theorems and the proofs. The student should be creating mathematics. Of course, he will actually be re-creating t with the aid of a teacher. He can be gotten to do this if ie is allowed and even encouraged to think intuitively, ut he cannot be expected to "discover" within the ramework of a logical development that is almost always highly sophisticated and artificial reconstruction of the original creative work. The constructive approach ensures understanding and teaches independent, productive hinking.

Teaching discovery is by no means a simple job. It calls for getting the students to use intuition, guessing, rial and error, generalization of known results, relating vhat is sought to known results, utilizing the geometrical neaning of algebraic statements, measuring, and dozens of other devices. It is relatively easy to get students to see that from

$$1 = 1$$
$$1 + 3 = 2^2$$
$$1 + 3 + 5 = 3^2$$

we can infer that

$$1 + 3 + 5 + \ldots + (2n - 1) = n^2;$$

in other words, the sum of the first n odd numbers is n^2. Likewise, a picture of an isosceles triangle suggests readily that the base angles are equal. Measuring the sides of several right triangles will lead students to the conclusion that the square of the hypotenuse equals the sum of the squares of the arms. But most often, teaching discovery calls for carefully preparing a series of simple questions which gradually lead to the desired conclusion. Even teaching discovery of such a relatively simple result as the quadratic formula is no longer easy.

The Socratic method of asking questions which lead students to discover a result must be used judiciously. The questions must be reasonable and answerable by most of the students. Otherwise the students will feel defeated and become disinterested. Students must acquire confidence in their own powers. They are more likely to do so if they contribute to the building of mathematics rather than being asked to learn a sophisticated theorem and proof which is the end result of much refashioning of older and cruder versions.

It is understandable and somewhat justifiable that authors of research papers do not communicate all the thinking, wandering, false starts, and guessing that led them to their theorems and proofs. It is deplorable if, by hiding from youngsters the existence of the fumbling and futile efforts, we give the impression that mathematicians reason directly and unfailingly to their conclusions. We not only rob students of the fun of discovery but we destroy the self-confidence that might have been built up if we had told them the truth and had led them to appreciate how difficult discovering the right theorem and proof really is.

Teachers are so anxious to cover ground that they hand down the final statements and proofs to the students, and since the students are not ready for such material they resort to memorization. To teach thinking we must let the students think, let the students build up the results and proofs even if incorrect. Let them learn also to judge correctness for themselves. Let's not push facts down students' throats. We are not packing articles in a trunk. This type of teaching dulls minds rather than sharpens them.

Culture is as much a process as a product. Until the sixteenth century it meant cultivation of the soil and, as we know, one does not put the fruit in the soil. One plants seeds and nourishes them. To teach students to pursue knowledge is part of a liberal education. We should get students to want to pursue it and not proclaim it to them.

The commonly accepted assertion that mathematics teaches people to think has not been tested. Mathematics instruction, old and new, has not been designed to teach people to think but to follow the leader, the teacher. In the traditional curriculum the students are taught to follow processes and repeat proofs. Today, under the new mathematics, the students memorize definitions and proofs. In fact, they are forced to memorize because the level of the material is beyond them.

In building mathematics constructively the genetic principle is enormously helpful as a guide. This principle says that the historical order is usually the right order and that the difficulties which mathematicians themselves experienced are just the difficulties our students will experience. Irrational numbers, negative numbers and complex numbers were bones in the throats of the best mathematicians. We may be sure, then, that the students are going to have trouble with these numbers. Hence we must be prepared for, and help them overcome, these particular difficulties, and we can be guided to a large extent by how mathematicians were convinced to accept and work with these numbers. Extending the distributive law to negative numbers will be of no help at all in getting students to feel at home with them.

Teaching constructively, as we have already remarked, is by no means easy. But there is no royal road. To enjoy the view from a mountain top one must get to the top. In mathematics there is no easy air-lift. The cables in young people's minds break down. But in the skillful employment of the process of discovery lies the true art of teaching. In this approach we arouse and develop the creative powers of the student and give him the delight of accomplishment.

Having gotten students to discover a result, by whatever means, one must face the matter of proof. There is no question that deductive proof is the hallmark of mathematics. No result is accepted into the body of mathematics until it has been proved deductively on the basis of an explicit set of axioms. However, proof as a criterion for the acceptance of a result by mathematicians and proof regarded from the standpoint of pedagogy are entirely different matters. We have already had occasion (Chapters 4 and 5) to point out the gross failings of insisting on logical or deductive proof as the pedagogical approach to mathematics. What is the alternative?

Contrary to what was done for generations in some branches of mathematics, such as Euclidean geometry where the teaching of deductive proof was the outstanding objective, and contrary to the approach of the modern mathematics curriculum, the basic approach to all new subject matter at all levels should be intuitive. This recommendation may appear to be treason to mathematics, but it is loyalty to pedagogy.

Admittedly the nature of intuition is somewhat vague. It denotes some direct grasp of the idea, whether it be a concept or proof. There may be a special intuitive faculty distinct from the logical faculty that criticizes and reasons. Whether or not there is an intuitive faculty, there are specific and explicit aids to the intuition which enable it to function. Primarily, one makes sense of mathematics through the senses, for, as Aristotle first put it, there is nothing in the intellect that was not first in the senses—though Leibniz added, except the intellect itself. Hence, one of the useful devices is a picture. Consider exhibiting several triangles to inculcate the *idea* as opposed to the definition: the union of three noncollinear points and the line segments joining them. Most students, even after being taught how to multiply $a + b$ by $a + b$, whether mechanically or logically, will state that $(a + b)^2 = a^2 + b^2$. A picture would help. It is clear from the picture (Fig. 11.1) that the area of a square whose side is $a + b$ is $a^2 + 2ab + b^2$.

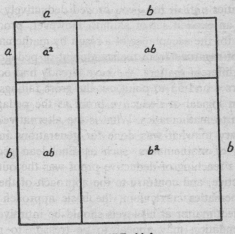

FIGURE 11.1

We include in the intuitive approach what are often called heuristic arguments. Through experience with actual objects a child can learn that $3 + 4 = 4 + 3$. The generalization that $a + b = b + a$ is heuristic.

Reasoning by analogy, though not at all deductive but heuristic, can be employed to great advantage. Students have great trouble working with irrational numbers expressed as radicals. Let us consider whether $\sqrt{2} + \sqrt{3} = \sqrt{5}$. The analogy might be $\sqrt{4} + \sqrt{9}$. Clearly this sum does not equal $\sqrt{13}$. And so we shall agree that $\sqrt{2} + \sqrt{3}$ does not equal $\sqrt{5}$. On the other hand, let us consider $\sqrt{2} \times \sqrt{3}$. Does this equal $\sqrt{6}$? The answer might be obtained by considering $\sqrt{4} \times \sqrt{9}$. This does equal $\sqrt{36}$. Therefore, we shall accept that $\sqrt{2} \times \sqrt{3} = \sqrt{6}$. As a matter of fact the Hindus and the Arabs, who were the first to work with radicals, reasoned entirely by analogy; and the Europeans, who learned these operations from the Arabs, did the same thing. The logical basis for irrational numbers was not erected until the late nineteenth century.

The intuition can be appealed to through physical arguments. Among the operations with negative numbers the multiplication of positive and negative numbers causes endless trouble. A well-known presentation based on gains and losses can convince students. Let's agree that if a man handles money a gain will be represented by a positive number and a loss by a negative number. Also, time in the future will be represented by a positive number and time in the past by a negative number. We can now use negative numbers to calculate the increase or decrease in a man's wealth. Thus, if he gains five dollars a day, three days in the future he will be fifteen dollars richer. In symbols $(+5)(+3) = 15$. If he loses five dollars a day, then three days in the future he will be fifteen dollars poorer. In symbols $(-5)(+3) = -15$. If he gains five dollars a day, then three days ago he was fifteen dollars poorer. In symbols $(+5)(-3) = -15$. Finally, if he loses five dollars a day, then three days ago he was fifteen dollars richer. In symbols $(-5)(-3) = 15$. Other situations such as water flowing in and out of a

ank may also be used to reinforce these rules of multiplication. Several texts use what is called the number line
and motions back and forth to teach the same rules.
Probably all of these situations should be used. Such concrete presentations may convince students that the definitions for multiplication with negative numbers are
reasonable and useful.

Another example of an appeal to physical happenings
would be to argue that if a ball is thrown up into the
air its velocity at the maximum height is zero because if
it were positive the ball would continue to rise whereas
if it were negative it would be falling. Such an argument
can be used to get students to set the velocity equal to 0
in order to calculate the time to reach maximum height.

All of the above devices, pictures, heuristic arguments,
induction, reasoning by analogy, and physical arguments are appeals to the intuition. Of course, the intuition is not static. Just as one's intuition about what to
expect in human behavior improves with experience so
does the mathematical intuition. The latter may indeed
suggest, as it did to Leibniz, that the derivative of a product of two functions is the product of the derivatives. The
conclusion should be tested, another heuristic measure,
and of course will be found to be false. Deeper analysis
will show that what holds for limits of functions does not
hold for derivatives, and the intuition will be sharpened
by this experience.

Clearly the intuitive approach can lead to error, but
committing errors and learning to check one's results are
part of the learning process. Truth, Francis Bacon noted,
emerges more readily from error than from confusion.
If the fear of errors is to be a deterrent, a child would
never learn to walk; and a student who makes no mistakes makes nothing else either.

The intuitive approach is further recommended in
that it is relatively easy to give a genuine or significant

motivation to a mathematical topic when this is introduced intuitively or heuristically because physical problems are the natural starting point for an intuitive approach. A logical presentation, on the other hand, is difficult to motivate because the latter is many stages removed from reality and is often artificial. How does one motivate a fraction when it is to be introduced as a set of equivalent ordered couples of natural numbers?

It is our contention that understanding is achieved intuitively and that the logical presentation is at best a subordinate and supplementary aid to learning and at worst a decided obstacle. Hence, instead of presenting mathematics as rigorously as possible, one should present it as intuitively as possible. As Professor Max M. Schiffer of Stanford University has stated it, "Never put logical carts before heuristic horses." Hermann Weyl defined the role of logic: "Logic is the hygiene which the mathematician practices to keep his ideas healthy and strong." And Jacques Hadamard, another of the famous mathematicians of our age, remarked that logic merely sanctions the conquests of the intuition. So far as understanding is concerned the use of logic in place of intuition amounts, in the words of the philosopher Arthur Schopenhauer, to cutting off one's legs in order to walk on crutches.

It is significant that when a mathematician reads a theorem which conflicts with his intuitive expectations his first move is to doubt not his intuition but the proof. He trusts his intuition more. If after having checked the proof carefully he becomes convinced that it is correct, he then inquires into what may be wrong with his intuition.

The intuitive approach can be strengthened immeasurably by incorporating in a mathematical classroom what is often called a mathematics laboratory. This would consist of apparatus of various sorts which can be used to demonstrate physical happenings from which mathematical results can be inferred. A few very simple laboratory devices have been designed and are used to a limited extent. One of these consists of (Cuisenaire) rods, which are no more than sticks of various lengths with which one can perform arithmetic operations with positive and negative integers. A second device called a geoboard (introduced by Caleb Gattegno) is a wooden board with regular rows and columns of nails. By stretching rubber bands over some of the nails one can form various geometric figures and demonstrate simple relationships. Another device calls for attaching objects of various weights to a spring and shows that the extension of the spring is proportional to the weight. This demonstration serves to introduce the linear function $y = kx$, wherein k depends on the tension in the spring. Still another simple device is the pendulum. One can increase or decrease the length of the pendulum and measure the period of the pendulum for various lengths. The goal is to have the students find the functional relationship between the length ℓ of the pendulum and the period T. The precise formula $T = (1/2\pi) \sqrt{\ell/32}$ is not readily obtained in this manner, but the numbers obtained by the student could be used as a basis for inferring, with the help of the teacher, what the precise formula is. Alternatively one could ask the students to find the formular for ℓ in terms of T; this is $\ell = 128\pi^2 T^2$, which is a little easier to discover, though again the teacher would have to aid in obtaining the precise formula. In either case the students would see the usefulness of a different kind of functional relationship. Too often students are introduced to a variety of functions without appreciating that different physical phenomena require different functions.

An excellent instrument to enliven and enrich the teaching of trigonometry is an oscilloscope. This is no more than a simplified television set. If we strike tuning forks of different frequencies more or less forcibly near a microphone, the oscilloscope displays on the screen the shapes of sinusoidal functions of various frequencies and amplitudes. These correspond to simple sounds. Moreover, by having students vocalize various sounds or by playing notes on several musical instruments the oscilloscope will show the graphs of these sounds. These graphs are readily shown to be combinations of sinusoidal graphs and so the student sees that vocal and musical sounds are no more than combinations of the simple sounds given off by tuning forks. Many more phenomena can be displayed on the screen of the oscilloscope. The main point, however, is to breathe life into the trigonometric functions. The student comes to appreciate that these dry, cold, "artificial" functions are omnipresent. He "speaks" trigonometry every time he utters a word. Moreover, he can readily be shown how this knowledge about sounds can be used in the design of the telephone, phonograph, radio, and other sound recording or reproducing instruments.

Laboratory material might be used by the teacher to perform demonstrations or be used by the students themselves working together in small groups. While the idea of a mathematics laboratory is not new it has not been used on a wide scale, nor has enough attention been paid to the invention of clever and helpful devices. This fine pedagogical aid has been neglected. The support of the collaboration of mathematics teachers and engineers to devise laboratory material, a project which has never been undertaken, would be money far more wisely spent than the tens of millions of dollars devoted to the development of the modern mathematics curriculum.

Does the reliance upon intuition, whether or not backed by physical demonstrations, mean that deductive proof and rigor will play no role in elementary mathematics education? Not at all. After a student has thoroughly understood a result and appreciates that the argument for it is merely plausible, the teacher can consider a deductive proof. However, the very idea of a deductive proof must be learned and this can be introduced only gradually. In no case should one *start* with the deductive approach, even after students have come to know what this means. The deductive proof is the final step. Moreover, the level of rigor must be suited to the level of the student's development. The proof need only convince the student. He should be allowed to accept and use any facts that are so obvious to him that he does not realize he is using them. The capacity to appreciate rigor is a function of the mathematical age of the student and not of the age of mathematics. This appreciation is acquired gradually and the student must have the same freedom to make intuitive leaps that the greatest mathematicians had. Proofs of whatever nature should be invoked only where the students think they are required. The proof is meaningful when it answers the student's doubts, when it proves what is not obvious. Intuition may fly the student to a conclusion but where doubt remains he may then be asked to call upon plodding logic to show the overland route to the same goal.

The level of rigor can, of course, be advanced as the student progresses. Poincaré makes this point. "On the other hand, when he is more advanced, when he becomes familiar with mathematical reasoning and his mind will be matured by this very experience, the doubts will be born of themselves and then your demonstration will be well received. It will awaken new doubts and the questions will arise successively to the child as they arose successively to our fathers to the point where only perfect rigor can satisfy him. It is not sufficient to doubt everything; it is necessary to know why one doubts." Rigor will not refine an intuition that has not been allowed to function freely. The student must experience the gradual passage from what he regards as obvious to the not-so-obvious and to the need for a fuller proof. He will discover the need for rigor rather than have it imposed on him.

This approach to rigor is more than a pedagogical concession. If one wishes to teach how mathematics developed and how mathematicians think, then the gradual imposition of rigor is precisely what does take place.

Apropos of this point E. H. Moore, in his article "On the Foundations of Mathematics," said, "The question arises whether the abstract mathematicians in making precise the metes and bounds of logic and the special deductive sciences are not losing sight of the evolutionary character of all life-processes, whether in the individual or in the race."

Critical thinking has been extolled as one of the great values to be derived from the study of mathematics. The modern mathematics leaders pride themselves on the fact that by emphasizing rigor they have promoted the development of critical thinking. But the capacity of the students to do critical thinking must be developed. If asked to assimilate and think critically about material that mathematicians took two thousand years to arrive at, the students will be overwhelmed—and instead of thinking will throw up their hands. To present young students with sophisticated mathematical formulations of basic ideas is entirely analogous to asking kindergarten students to be critical of a work in philosophy. There is no short-cut to the development of the critical faculty. As E. H. Moore put it, "Sufficient unto the day is the rigor thereof." And by the day he meant the student's age.

Fortunately, young people will accept as rigorous and acquire a feeling for proof from proofs that are really not rigorous. Is this deception? No! It is pedagogy. At any rate, it is no more deception than we practice on ourselves. As our own capacity to appreciate more rigorous proofs increases, we are able to see flaws in the cruder proofs taught to us and to master sounder proofs. This is also how the great mathematicians gradually improved the rigor of their subject. But let us not forget that there are no final rigorous proofs. Not all the symbolism of modern symbolic logic, Boolean algebra, set theory, and axiomatic methods have made or can make mathematics perfectly rigorous.

With respect to the technique of presentation there are additional principles to be observed. In place of abstract concepts we should as far as possible present concrete examples. Thus it does not matter if a student cannot give a general definition of a function. It suffices if he knows concrete functions such as $y = 2x$ and $y = x^2$ and learns how to work with them. After some experience with functions the student will be able to make his own definition. And if after further experience the definition has to be modified, no calamity has occurred. This is precisely how the mathematicians proceeded in the years from 1700 to 1900. Again it does not matter whether a student can define a polygon as long as he can recognize and work with it. In this connection a picture is worth a thousand words. We know what dogs and men are and we distinguish the two successfully without being able to define either one. Piaget has pointed out that young people need to build up layers of experience before they can master abstractions. Insight into all kinds of knowledge comes and grows only with experience. As Whitehead has put it, "There is no royal road to learning through an airy path of brilliant generalizations. . . . The problem of education is to make the pupil see the wood by means of the trees."

Instead of multiplying terminology we should introduce as few terms as possible. Common words, preferably those already familiar to the student, should be used even though the words are given technical meaning. Terminology should be kept to a minimum. Verbalization comes after understanding, and it can be the student's verbalization rather than the artificial compressed language of modern mathematics.

As in the case of terminology, symbolism too should be kept to a minimum. Symbols scare students. Moreover, the meaning of a symbol must be remembered and so is often more of a burden than an aid. The gain in brevity may not compensate for the disadvantages.

A few words about content may be in order. The two considerations previously discussed, the need to offer a liberal education and the need to motivate youngsters, should have the highest priority in determining content on the elementary and high school levels. It would of course be desirable, in view of the somewhat sequential nature of mathematics, to incorporate the subjects that are normally taught at the various levels so that those students who pursue the subject further at the college level are not seriously delayed by the omission of necessary subject matter. Fortunately, it is possible to teach most of the standard material of the traditional curriculum with the proper motivation and exposition of its significance. Though fortunate, this fact is not fortuitous. The standard material, except for a few topics and possible reordering in algebra, is the material that has been found useful—and this is why it has been taught in preference to many other possible topics. However, one should not hesitate to depart from some of it if a richer and more vital course can be fashioned. For example, statistics and the use of computers are substantial topics and may arouse interest. Any topics that may have to be omitted and that are necessary to the further pursuit of mathematics can be incorporated in courses addressed to students who are definitely committed to mathematics.

Beyond these considerations there is the matter of priority in importance for mathematics and science.

There is nothing intrinsically wrong with set theory and in fact it is essential at the *advanced* undergraduate and graduate levels. But it does not warrant time on the elementary and high school levels. Arithmetic, algebra and geometry are far more important, and set theory does not contribute to the learning of these subjects. Analogous remarks apply to Boolean algebra, congruences, symbolic logic, matrices, and abstract algebra.

Many advocates of modern mathematics have cut down drastically on Euclidean geometry. Indeed, the usual modern text replaces much of synthetic geometry by analytic geometry. Some extreme modernists have favored abolishing all synthetic geometry. "Down with Euclid" and "Euclid must go" have appeared as slogans in the new mathematics movement. Such a step would be tragic. Not only is synthetic geometry an essential part of mathematics in which Euclidean geometry is the base, but geometry furnishes the pictorial interpretation of much analytic work. Mathematicians usually think in terms of pictures, and geometry not only furnishes the pictures but suggests new analytical theorems. It is incredible that knowledgeable mathematicians should seek to oust synthetic geometry.

There is a widely known story that a mathematician lecturing before a class got stuck in the middle of a proof. He went over to the side of the board, drew a few pictures, erased them and was then able to continue his lecture. What this story implies about pedagogy is seriously disturbing but it does speak for the uses of pictures.

As far as mathematical content is concerned all of the desirable changes amount to no more than minor modifications of the traditional curriculum, and all talk about modern society requiring a totally new kind of mathematics is sheer nonsense.

To delineate the approach to and mathematical content of courses is not enough. The concentration on curriculum has been to a large extent an escape from reality. The bigger and more vital problem is the education of teachers. Since the curriculum must furnish a liberal education and above all supply motivation for the subjects and topics we do teach, we shall have to introduce, respect and remunerate a new class of professors, mathematical scholars, who can offer the proper training of teachers. The traditional curriculum was fashioned by relatively uninformed mathematicians with no pedagogical insight. The modern mathematics curriculum was fashioned jointly by such people and by narrow researchers in pure mathematics with as little pedagogical insight. The people we need will have to possess breadth not only in mathematics but also in the various areas in which mathematics has influenced our culture. They will also have to be educators. This means that they will have to know how much young people can handle of abstractions and proofs, and which motivations will appeal to a ten-year-old and which to a fourteen-year-old. Moreover, the breadth and openness of mind desired of the ideal scholar would require that he also see mathematics from the nonmathematician's point of view so that he can appreciate the attitudes and problems of young people. To put the matter crudely, the proper mathematical scholar must not only know his stuff but also know whom he is stuffing. We need, in other words, professors of broad scholarship and educational insight as opposed to the self-centered, narrow researcher.

It is most likely that the type of person we need will have to be trained by the mathematics departments of graduate schools and be attached to the universities. The proper program for such people does not exist at present. Nor, unfortunately, does it seem likely that the academic graduate schools will readily undertake such training. The inertia and narrowness of the graduate schools may be seen from a closely related problem. The graduate

chools train Ph.D.'s for research. However, it has been
ecognized for some ten or fifteen years that most of the
*h.D.'s trained by the graduate schools, as many as
eventy-five or eighty per cent, do not do research after
*btaining the Ph.D. These people take positions in the
our-year colleges and smaller universities where teaching
s the main activity and concern. Hence ten years ago a
oint committee of the American Mathematical Society
nd the Mathematical Association of America recom-
nended that the graduate schools offer an alternative
*rogram that would be directed toward the preparation
*f college teachers rather than toward the training of
esearchers, toward breadth rather than depth. These
*rospective teachers could be awarded the usual doc-
*or's degree or a new degree to be called the Doctor of
*rts. Realistic and wise as this recommendation seems
o be, no major graduate school of the country took it up.
*s of about 1970 ten of the less prestigious universities
*egan to experiment with such a program under the
*upport of the Carnegie Corporation. Experience with
*niversity administrations leads one to wonder whether
he wisdom of the program or the financial arrangement
*as the greater inducement to experiment.

The professors are not interested in training college
*eachers. They regard such work as demeaning and as
*owering the quality of their departments. To train
mathematicians who would have the breadth and compe-
*ence to treat pedagogical problems of the elementary
and high schools calls for a still wider departure from
the research-oriented programs, and the present professors will not and in fact are not prepared to direct
such training.

Some universities have tried to meet the problem of
developing better schoolteachers by selecting mathematics professors as professors of education. This idea,
intrinsically sound, does not work—and the reasons are

pertinent. Mathematicians have looked down on mathematics education as an inferior activity (in the past justified by the low quality of the schools of education). Hence, those mathematicians who are comfortable about their roles in mathematics departments will not accept positions as professors of mathematics education. Unfortunately, the universities that made the move to hire mathematicians as mathematics education professors sought mathematicians with prestige, and such men are all the more reluctant to accept what their colleagues would regard as an inferior position. It was often the case that those who were attracted to becoming mathematics education professors were either not particularly successful in their role as mathematicians or were attracted by factors such as money or the greater prestige of the university to which they would be going. But such men would not necessarily qualify as educators, and in fact they are not. The best choices for mathematics education professors would be broadly educated mathematicians with a genuine interest in education, but such people do not stand out in the mathematical world and would be harder to locate. Nor would their names add prestige to the university that invites them, because prestige in mathematics is built upon research and in view of today's intense specialization this almost always means narrowness. Thus in the present university atmosphere the endeavor to get mathematicians to serve as education professors either attempts to mate a horse and a donkey (there is no implication as to which is which) and produces a sterile progeny, or it succeeds in attracting men who fall between two stools and fall hard.

The primary function of the mathematical scholars would be to improve the education of elementary school and high school teachers of mathematics. At present the knowledge of mathematics which these teachers possess is often inadequate; nor are they required to know anything about the uses and cultural reaches of mathematics. In particular they know no science. Clearly such teachers are not prepared to teach a liberal arts course in mathematics, to motivate mathematics through non-mathematical problems, or to apply mathematics. Most mathematics teachers have been assured that mathematics is important and they tell this to their students. However, they are unable to show how it is important, and so their attempts to convince students lack conviction. Students can see through hollow assurances.

Other forces also operate against the possibility of training teachers properly. The Committee on the Undergraduate Program in Mathematics (CUPM), a committee of the Mathematical Association of America, has prepared a Course Guide for the Training of Junior High and High School Teachers of Mathematics. The college courses recommended for this training are analytical geometry and calculus, abstract algebra, linear algebra, geometry, probability and statistics, logic and sets. There was not even the suggestion, much less a requirement, that these prospective teachers should study science.

At the present time the schoolteacher is in a dilemma. Many are located in small communities which are not close to a major university. Those who could conveniently take mathematics courses at a university are not much better off. If they go to the graduate division of the university they must undertake a masters or doctoral program in mathematics. The courses in these programs are directed to prospective research mathematicians and so do not offer the breadth that schoolteachers need. The courses are also too difficult. The alternative for the teachers is to go to a School of Education. Here they may learn about education but not subject matter. So the teachers are in no better position to judge what is important in mathematics and to develop the competence to teach and write texts for the schools.

The training of good teachers is far more important than the curriculum. Such teachers can do wonders with any curriculum. Witness the number of good mathematicians we have trained under the traditional curriculum, which is decidedly unsatisfactory. A poor teacher and a good curriculum will teach poorly whereas a good teacher will overcome the deficiencies of any curriculum.

Who is to fashion the "right" curriculum of the future? The broad mathematical scholars and the experienced, mature, well-educated elementary and high school teachers are the proper people. Research people, psychologists, and education professors of the current type may serve as consultants but certainly should not lead this work. Beyond this, the schoolteachers should be the arbiters of what is to be taught and how it is to be taught. They are the ones who have worked with young people and know best what motivates them and what degree of abstraction they can absorb.

What criterion of success should we utilize? It should not be how far students have advanced in mathematics —many high schools boast today that their students take calculus—nor what sophisticated notions they have been taught. When we reach the stage where fifty per cent of the high school graduates can honestly say that they like mathematics and appreciate its significance, then we shall have attained a large measure of success in the teaching of mathematics.

In view of the shameful record of mathematics education past and present, how is it that mathematics has survived and that we have a flourishing, if not altogether sound, mathematical activity in this country? I think that we owe what we have accomplished to a few wise, mature, devoted teachers who by their care in choosing what to emphasize and by their personal charm and magnetism have attracted some students to mathematics. Those noble souls have saved us from disaster.

Bibliography

Aichele, Douglas B. and Robert E. Reys: *Readings in Secondary School Mathematics*, Prindle, Weber and Schmidt, Inc., Boston, 1971.

Allendoerfer, C. B.: "The Narrow Mathematician," *American Mathematical Monthly*, 69, 1962, 461–469.

Begle, E. G.: "Open Letter to the Mathematical Community," *The Mathematics Teacher*, 59, 1966, 341 and 393. Also in *Science*, 151, 1966, 632.

Birkhoff, George D.: "The Mathematical Nature of Physical Theories," *American Scientist*, 31, 1943, 281–310.

Cambridge Conference on School Mathematics: *Goals for School Mathematics*, Houghton Mifflin Co., Boston, 1963.

Carrier, George F. et al: *Applied Mathematics: What is Needed in Research and Education*, SIAM Review, 4, 1962, 297–320.

Council for Basic Education: *Five Views of the "New Math,"* Washington, D.C., April 1965. Occasional Papers #8. Articles by H. M. Bacon, A. Calandra, R. B. Davis, M. Kline, E. E. Moise.

Deans, Edwina: *Elementary School Mathematics: New Directions*, United States Office of Education, 1963.

DeMott, Benjamin: "The Math Wars," *The American Scholar*, Spring Issue, 1962, 296–310.

Fehr, Howard F.: "The Role of Physics in the Teaching of Mathematics," *International Symposium on the Coordination of Instruction in Mathematics and Physics,* Belgrade, 1962, 25–32.

Fehr, Howard F.: "Sense and Nonsense in a Modern School Mathematics Program," *The Arithmetic Teacher,* Feb. 1966, 83–91.

Fehr, Howard F. and J. Fey: "The Secondary School Mathematics Curriculum Study," *American Mathematical Monthly,* 76, 1969, 1132–1137.

Feynman, Richard P.: "New Textbooks for the New Mathematics," *Engineering and Science,* 28, 1965, 9–15.

Fremont, Herbert: "New Mathematics and Old Dilemmas," *The Mathematics Teacher,* 60, 1967, 715–719.

Fremont, Herbert: *How to Teach Mathematics in Secondary Schools,* W. B. Saunders Co., Philadelphia, 1969.

Hammersley, J. M.: "On the Enfeeblement of Mathematical Skills by 'Modern Mathematics' and by Similar Soft Intellectual Trash in Schools and Universities," *Bulletin of the Institute of Mathematics and Its Applications,* Oct. 1968, 66–85.

Heath, R., ed.: *New Curricula,* Harper and Row, 1964.

Herstein, I. N.: "On the Ph.D. in Mathematics," *American Mathematical Monthly,* 76, 1969, 818–824.

Khinchin, A. Ya.: *The Teaching of Mathematics,* American Elsevier Publishing Company, Inc., N.Y., 1968.

Kline, Morris: *Mathematics and the Physical World,* T. Y. Crowell Co., N.Y., 1959.

Kline, Morris: *Mathematics, A Cultural Approach,* Addison-Wesley Publishing Co., Reading, Mass., 1962.

Kline, Morris: "Mathematical Texts and Teachers: A Tirade," *The Mathematics Teacher,* 49, 1956, 162–172.

Kline, Morris: "The Ancients Versus the Moderns: A New Battle of the Books," *The Mathematics Teacher,* 51, 1958, 418–427.

Kline, Morris: "A Proposal for the High School Mathematics Curriculum," *The Mathematics Teacher,* 59, 1966, 322–330.

Kline, Morris: "Intellectuals and the Schools: A Case History," *The Harvard Educational Review*, 36, 1966, 505–511.

Kline, Morris: "Logic Versus Pedagogy," *American Mathematical Monthly*, 77, 1970, 264–282.

Manheimer, Wallace: "Some Heretical Thoughts from an Orthodox Teacher," *The Mathematics Teacher*, 53, 1960, 22–26.

McIntosh, Jerry A., ed.: *Perspectives on Secondary Mathematics Education*, Prentice-Hall, N.J., 1971.

Merriell, D. M.: "Second Thoughts on Modernizing the Curriculum," *American Mathematical Monthly*, 67, 1960, 76–78.

Minsky, Marvin: "Form and Content in Computer Science," *Journal of the Association for Computing Machinery*, 17, 1970, 197–215.

Moise, Edwin: "Some Reflections on the Teaching of Area and Volume," *American Mathematical Monthly*, 70, 1963, 459–466.

Moore, E. H.: "On the Foundations of Mathematics," *American Mathematical Society Bulletin*, 9, 1903, 402–424.

National Council of Teachers of Mathematics: *The Revolution in School Mathematics*, National Council of Teachers of Mathematics, Washington, D.C., 1961.

National Council of Teachers of Mathematics: *An Analysis of New Mathematics Programs*, National Council of Teachers of Mathematics, Washington, D.C., 1963.

National Council of Teachers of Mathematics: "The Secondary Mathematics Curriculum," *The Mathematics Teacher*, 52, May 1959, 389–417.

Neumann, John von: "The Mathematician," an essay in E. B. Heywood, ed.: *The Works of the Mind*, University of Chicago Press, 1957.

Nevanlinna, Rolf: "Reform in Teaching Mathematics," *American Mathematical Monthly*, 73, 1966, 451–464.

Sawyer, W. W.: "The Reconstruction of Mathematical Education," *Journal of Engineering Education*, 51, 1960, 98–113.

Smith, D. B.: "Some Observations on Mathematics Curriculum Trends," *The Mathematics Teacher,* 53, 1960, 85–89.

Stone, Marshall H.: "The Revolution in Mathematics," *American Mathematical Monthly,* 68, 1961, 715–734; also in *Liberal Education,* 47, 1961, 304–327.

Stoker, James J.: "Some Observations on Continuum Mechanics," *American Mathematical Society Bulletin,* 68, 1962, 239–278.

Synge, John L.: "Focal Properties of Optical and Electromagnetic Systems," *American Mathematical Monthly,* 51, 1944, 185–187.

Thom, René: " 'Modern' Mathematics: An Educational and Philosophical Error?" *American Scientist,* 59, 1971, 695–699.

Weinberg, Alvin M.: "But Is the Teacher Also a Citizen?" *Science,* 149, August 6, 1965, 601–606.

Weinberg, Alvin M.: *Reflections on Big Science,* M.I.T. Press, 1967. See especially p. 160.

Whitehead, Alfred North: *The Aims of Education and Other Essays,* The New American Library, N.Y., 1949.

Whitehead, Alfred North: *Essays in Science and Philosophy,* Philosophical Library, N.Y., 1948.

Wittenberg, Alexander: "Sampling a Mathematical Sample Text," *American Mathematical Monthly,* 70, April 1963, 452–459.

MORRIS KLINE did his undergraduate work at New York University and received a Ph.D. there in mathematics. He engaged in post-doctoral research at the Institute for Advanced Study in Princeton, spent a year in Germany as a Guggenheim Fellow, and was Director of the Division of Electromagnetic Research at NYU's Courant Institute of Mathematical Sciences for twenty years. He is presently Professor of Mathematics at NYU. Among Professor Kline's books are *Mathematics in Western Culture* and *Mathematics and the Physical World*. Professor and Mrs. Kline have three children. They live in Brooklyn, New York.